CLEANROOM CLOTHING SYSTEMS

People as a Contamination Source

Bengt Ljungqvist and Berit Reinmüller

PDA
Bethesda, MD, USA
DHI Publishing, LLC
River Grove, IL, USA

10 9 8 7 6 5 4 3 2 1

ISBN: 1-930114-60-5
Copyright © 2004 Bengt Ljungqvist and Berit Reinmüller. All rights reserved.

PDA
3 Bethesda Metro Center
Suite 1500
Bethesda, MD 20814
United States
301-986-0293

Davis Healthcare International Publishing, LLC
2636 West Street
River Grove
IL 60171
United States
www.DHIBooks..com

AUTHOR BIOGRAPHIES

Dr. B. Ljungqvist received his Ph.D. in 1978 and was appointed Professor of Safety Ventilation in 1986 at Kungl Tekniska Högskolan, Stockholm. Since 1978 he has worked in the field of safety ventilation, both as a scientist and as a consultant. In 1993 he was appointed Visiting Professor in the Department of Environmental Sciences and Engineering at the University of North Carolina at Chapel Hill, USA. He is a past Chairman of the Nordic Association of Contamination Control (R3-Nordic). As a consultant he is contracted by pharmaceutical companies, hospitals, laboratories, etc., to solve problems within this field.

Dr. B. Reinmüller received her Ph.D. in 2001. She has spent 30 years in the pharmaceutical manufacturing field specializing in the areas of contamination control, environmental monitoring, validation, and microbiological risk assessment. She is also representing Sweden in the ISO 209 work on "Cleanrooms and associated controlled environments". She is employed as Senior Researcher in Building Services Engineering at Kungl Tekniska Högskolan in Stockholm, and also contracted as a consultant for cleanroom industry.

CONTENTS

Contents

PREFACE

Today clothing and clothing systems for cleanrooms are mainly tested with regard to material properties, such as particle generation, particle filtration and resistance to wear. The dispersal chamber or "body-box" has been used for studying the protective efficacy of clothing systems in use. A modified dispersal chamber has recently been installed at KTH (Royal Institute of Technology, Stockholm). Tests and comparative studies have been performed in the dispersal chamber on selected clothing systems.

The results are given in total number of airborne particles (≥ 0.5 μm) per cubic meter and airborne aerobic colony forming units (CFU) per cubic meter. Statistical evaluations of the results have been performed. The source strengths of the contamination source people wearing modern cleanroom clothing systems have been estimated.

The work presented here is partly based on publications from scientific theses, the PDA Journal of Pharmaceutical Science and Technology and the European Journal of Parenteral & Pharmaceutical Sciences (Reinmüller, 2001; Reinmüller and Ljungqvist, 2002, 2003; Ljungqvist and Reinmüller, 2003a, 2003b).

The authors wish to thank the Swedish Council for Textile Research, Berendsen Textile Service, Pharmacia Corporation and Ferring AB, which have financially supported the project. During the project Labora AB and AB NINOLAB have given their support in the form of instrument calibration and disposable material necessary for the studies.

Stockholm
December 2003

Bengt Ljungqvist
Berit Reinmüller

INTRODUCTION

People disperse fragments from the skin and the airborne dispersion will vary from person to person and from time to time. The prime function of cleanroom clothing is as a filter protecting product and processes from airborne contamination. Clothing systems should be designed to envelope a person and not allow significant amounts of contaminants to be dispersed into the cleanroom. Properties of the fabrics used for cleanroom clothing can be assessed by measurements of air permeability, particle retention and pore size. The fabric itself should disperse the minimum of particles and be resistant to breakdown and tearing. However, the effectiveness of cleanroom clothing will deteriorate due to factors such as aging, wear, washing, drying and sterilization.

Today, clothing and clothing systems for cleanrooms and associated controlled environments are mainly tested with regard to material properties such as particle generation, particle filtration, and resistance to wear-related damage.

The combined filtration efficacy of fabric, construction and design of the clothing can be evaluated in a dispersal chamber or "body box". The dispersal chamber has been used for studying the protective efficacy of clothing systems in use (Whyte et al., 1976; Hoborn, 1981; Whyte and Bailey, 1985).

The increasing cleanliness demands in a variety of industries require in-depth knowledge regarding the performance of modern cleanroom clothing systems.

OBSERVATIONS FROM CASE STUDIES

Over the years, studies have been performed in the pharmaceutical industry on the influence of clothing systems and the microbiological contamination level in an aseptic environment. One of the authors (Reinmüller, 2001) has conducted an investigation, in an aseptic filling room, in which comparative measurements were carried out between two types of cleanroom coverall. The addition of long boots to the coverall was also evaluated.

The single test subject performed a series of movements simulating typical working activities in an aseptic cleanroom of mixed air flow type. During the movements, the concentration of airborne particles (≥ 0.5 μm and ≥ 5 μm) and airborne CFU was measured. Airborne particles were counted by a particle counter (DPC, HiacRoyco) and airborne CFUs measured by an RCS Air Sampler. Both the RCS Air Sampler and the measuring probe of the particle counter were placed approximately two meters above the floor. Two types of cleanroom coveralls with separate hoods were used; both garments were made of 100% polyester fabric. A newly introduced coverall using a thinner variety of fiber and more densely woven fabric was tested along with a second coverall made from thicker fibers. The effect of knee-length boots compared to normal cleanroom shoes was also evaluated. Testing was done on each coverall type in conjunction with knee-length boots and with cleanroom shoes. Table 1 shows the measured particle levels; each value represents a mean value of five sampling periods.

1

Table 1. The number of airborne particles equal to and larger than 0.5 μm and 5 μm, respectively per cubic meter during comparison of cleanroom clothing systems. Measurements performed two meters above the floor.

Type of clothing system	Without long boots Number of particles/m^3		With long boots Number of particles/m^3	
	$\geq 0.5\ \mu$m	$\geq 5\ \mu$m	$\geq 0.5\ \mu$m	$\geq 5\ \mu$m
Old (thick fiber)	159,250	4,025	14,700	175
Old (thick fiber)	240,450	4,550	16,800	315
New (thin fiber)	105,350	1,960	12,950	945
New (thin fiber)	218,050	5,565	24,150	420
Mean Value	180,775	4,025	17,150	464

Table 1 shows a reduction of airborne particles of approximately 90% when cleanroom coverall and hood were used together with knee-length tested boots. No significant difference was found between the two compared fabrics in the study. Airborne CFU were measured at the same location as airborne particles and showed a similar reduction in the number of aerobic CFU when knee-length boots were used. To investigate this improvement over a long period of time, environmental monitoring results from two similar filling rooms were compared. The size of the rooms, the number of air changes, and the number of people working within the room was in the same range. The aseptic operations performed within the rooms were comparable. The main difference between the two environments was that operators were using knee-length boots together with the cleanroom coverall in one room and not using knee-length boots with the cleanroom coverall in the other room. Table 2 shows a summary of the microbiological environmental monitoring results for a period of six months.

Table 2 shows that the cleanroom clothing of operators has an influence on the microbiological burden in the room. Not only is the number of airborne CFU affected, but also the number of CFU found on horizontal surfaces and floors within the rooms. Knowledge about cleanroom clothing systems with regard to the protection efficacy against airborne particles (including CFU) is of importance in the choice of adequate clothing systems to be used during pharmaceutical manufacturing.

Table 2. Summary of environmental microbiological results from two aseptic filling rooms measured during activity over a period of six months.

Type of samples	Clothing with/without long boots	Number of samples	Result in %		
			Good*	Normal**	Above Normal***
Air, active sampling	With	107	97	3	0
Horizontal surfaces, rodac plates	With	34	100	0	0
Floors, rodac plates	With	21	100	0	0
Air, active sampling	Without	126	79	15	6
Horizontal surfaces, rodac plates	Without	54	89	9	2
Floors, rodac plates	Without	45	80	13	7

* **Good:** Number of detected CFU between 0 CFU and half of the normal level for each kind of sample (air, surface and floor).

** **Normal:** Number of detected CFU from half of the normal level and up to normal (alert level) for each kind of sample (air, surface and floor).

*** **Above Normal:** Number of detected CFU above the normal level (alert level) for each kind of sample (air, surface and floor); includes results above action level.

METHODS

Today, clothing and clothing systems for cleanrooms and associated controlled environments are mainly tested with regard to material properties such as particle generation, particle filtration, and resistance to wear and tear-related damage. A dispersal chamber or "body-box" has been used to study cleanroom garment protection efficiency by, e.g., Whyte et al. (1976), Hoborn (1981) and Whyte and Bailey (1985). Measurements are carried out in order to relate airborne dispersal of total particulates and/or viable particulates to the quality of fabrics and the design of the cleanroom clothing. The increasing cleanliness demands in a variety of industries require in-depth knowledge regarding the performance of modern cleanroom clothing systems.

A specially designed dispersal chamber with HEPA-filtered supply air and separate exhaust air has been qualified for the evaluation of cleanroom clothing systems (Ytterman, 1998). The arrangement is shown schematically in Figure 1. The vertical unidirectional air velocity is adjustable from 0.1 m/s to 0.6 m/s and the dispersal chamber is pressurized relative to the adjacent area. Temperature and relative humidity are not controlled, but have during the tests been in the range 20–26 °C and 25–55% RH, respectively.

Containment tests in the dispersal chamber were carried out to evaluate different clothing systems by measuring the concentration of airborne total particulates and viable particulates (as aerobic CFUs) in the exhaust air. The total number of airborne particles was determined using a particle counter (DPC; HiacRoyco 245) and viable particles collected primarily using a slit-sampler (brand name FH3) and in some cases additionally using a sieve-sampler (brand name Andersen 6-Stage Sampler). All instruments were operated according to the manufacturers' instructions. Microbiological growth medium for all tests was standard medium Tryptic Soy Agar (TSA) in 9 cm Petri dishes. The TSA plates were incubated for not less than three days at 32 °C followed by not less than two days at room temperature. The recorded CFU were characterized by phase contrast direct microscopy.

Figure 1. Principal arrangement of dispersal chamber (body-box).

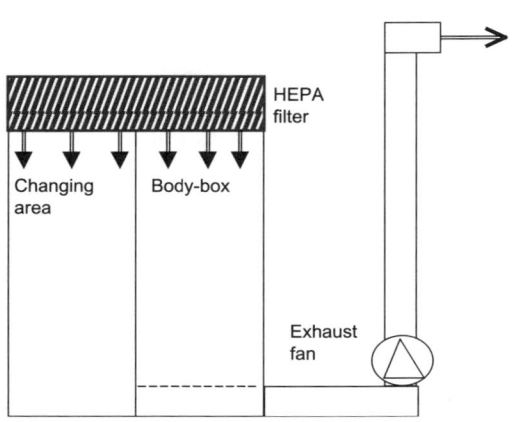

During the measurements the test subjects (young men) performed standardized cycles of movements that included head turns, arm movements, walking in place, and knee bends. These movements, are in principle, comparable with those described in IES-RP-CC003.2 (1993). Prior to each new cycle of movement, the test subject stood still to avoid the influence of particle generation from the previous test cycle.

The tests were performed on different occasions and will be referred to as Test Series 1, Test Series 2 and Test Series 3. In Test Series 1, only one test subject was used throughout and each cycle of movements was performed over one minute for each movement type. In Test Series 2 there were ten different test subjects. Each cycle of movement was extended from one to three minutes. The cycle consisting of head turns was not used in Test Series 2, as this activity did not result in a significant release of airborne total particulates and viable particulates. In Test Series 3 each of five test subjects performed six test sequences with used, modern cleanroom clothing systems with variations in fabrics and, as a comparison, two sequences with surgical clothing systems. The used clothing systems had passed through 25 and 50 washing/sterilizing cycles, respectively.

TEST SERIES 1

Results from Test Series 1 have been reported by Reinmüller and Ljungqvist (2000). Figures 2 and 3 illustrate results obtained from three different clothing systems. These are items of reference clothing referred to as Ref. 1 and 2, disposable coats referred to as Disp. 1 and 2, and complete cleanroom clothing set referred to as Cleanroom Clothing 1 and 2. The reference clothing consisted of street clothing (jeans and T-shirt). The disposable coat was a long sleeve coat with cuffs at the neck and arms and tied at the back. The cleanroom clothing in this test had been used, washed and autoclaved several times. Figure 2 shows the number of airborne particles (≥ 0.5 µm) per cubic meter and Figure 3 shows the number of airborne CFU per cubic meter.

Figure 2. Number of airborne total particles ≥ 0.5 µm per cubic meter measured from tests with three different clothing systems (one test subject).

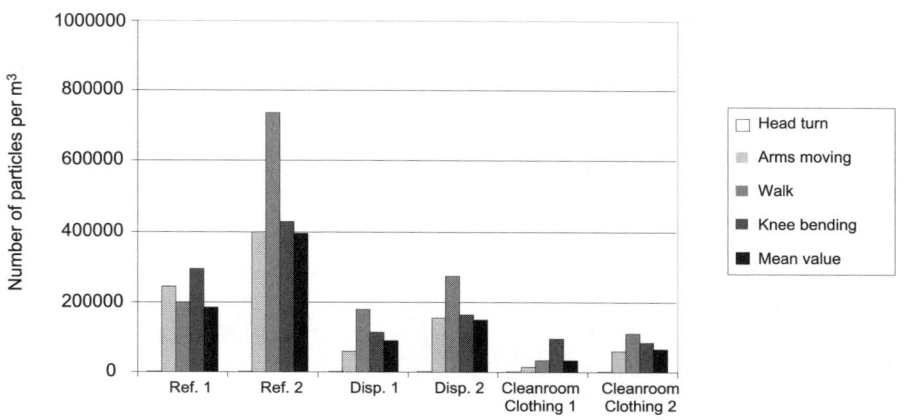

Figure 3. Number of airborne CFU per cubic meter measured during the same tests shown in Figure 2 (one test subject).

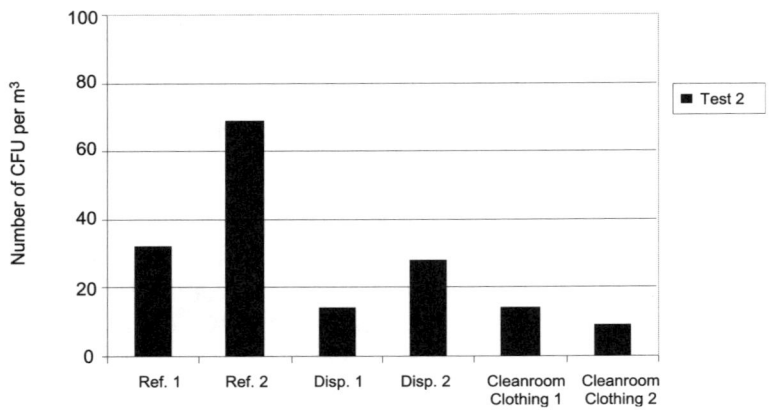

Table 3. Ratios based upon mean values of the total number of airborne particles (≥ 0.5 μm and ≥ 5 μm) per cubic meter, to the number of airborne aerobic CFU per cubic meter.

Test 2	Ratio of No. of particles ≥ 0.5 μm to No of CFU	Ratio of No. of particles ≥ 5 μm to No of CFU
Reference 1	5.7×10^3	415
Reference 2	5.7×10^3	411
Disposable coat 1	6.3×10^3	454
Disposable coat 2	5.3×10^3	319
Cleanroom clothing 1	2.7×10^3	68
Cleanroom clothing 2	7.0×10^3	192

Ratios based upon mean values of the total number of airborne particles per cubic meter equal to and larger than 0.5 μm and equal to and larger than 5 μm, respectively, to the number of aerobic CFU per cubic meter are given in Table 3. The results presented in Table 3 show a relationship between total number of airborne particles and number of airborne aerobic CFU. Figure 4 shows the number of airborne CFU per cubic metre in relation to the number of washing/sterilizing cycles of the tested cleanroom clothing systems.

Figure 4. Number of airborne CFU per cubic meter in relation to the number of washing/sterilizing cycles of the cleanroom clothing system (one test subject).

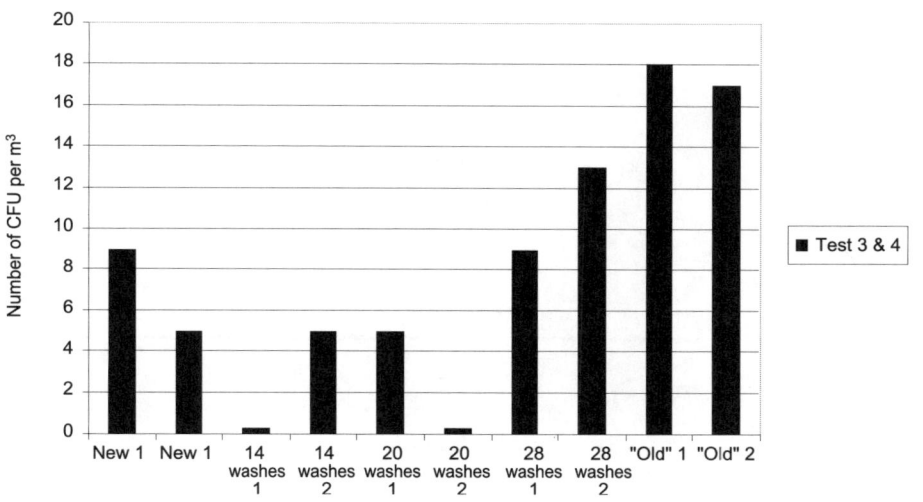

The data given in Figure 4 demonstrate the effect of washing/sterilizing cycles on cleanroom garment performance. The results indicate that the cleanroom garment performance deteriorates markedly as a result of the number of washing/sterilizing cycles. The data suggest that there are a limited number of washing/sterilizing cycles that a cleanroom garment can go through for an acceptable level of performance. It indicates that this limit is lower than the 50–60 washing/sterilizing cycles of the "old" garments in Figure 4. However, more tests are necessary if a general limit is to be established.

During the same test sequence shown in Figure 4, the total number of airborne particles per cubic meter has been measured for particles equal to and larger than 0.5 μm and particles equal to and larger than 5 μm, respectively. Ratios based upon the mean values of the total number of airborne particles per cubic meter to the number of aerobic CFU per cubic meter are given in Table 4.

Table 4. Ratios based upon the mean values of the total number of airborne particles (≥ 0.5 µm and ≥ 5 µm) per cubic meter, to the number of airborne aerobic CFU per cubic meter.

Number of Washes	Ratio of No of particles ≥ 0.5 µm to No. of CFU	Ratio of No of particles ≥ 5 µm to No. of CFU
New clothing –1	0.6×10^3	11
New clothing –2	1.3×10^3	40
14 Washes –1		
14 Washes –2	6.3×10^3	168
20 Washes –1	3.7×10^3	79
20 Washes –2		
28 Washes –1	1.6×10^3	39
28 Washes –2	1.3×10^3	39
"Old" –1	1.0×10^3	34
"Old" –2	1.1×10^3	46

The results indicate a relationship between the number of total airborne particles and the number of aerobic CFU when people are the contamination source. The ratio between particles equal to and larger than 0.5 µm and airborne aerobic CFU in this instance is estimated to be in the range between 600 to 1 and 7,000 to 1. It should be noted that the measuring equipment in Test Series 1 used different sampling durations for total particulates and for CFUs; this is why the above-given range should be seen only as approximate. A relationship has also been established from studies of cleanroom environments; see Ljungqvist and Reinmüller (1995, 1998).

The results in Test Series 1, with only one test subject, should be considered as indicative. However, the results show the relevance of the dispersal chamber tests in the evaluation of modern cleanroom clothing and clothing systems.

Analyses of the data show that head turn movement contributes only a negligible amount to the generation of particles and CFUs; this is why the head turns are excluded in the following studies.

TEST SERIES 2

In Test Series 2, as mentioned earlier, the standardized cycles of movements were arm movements, knee bends, and walking in place. A test sequence consists of each of the three cycles of movements done for three minutes each and repeated twice with a rest of one minute between, which equals 18 minutes of activity in total. In these tests, the sampling duration of total particles and CFUs coincided during all measurements. The cleanroom clothing systems were washed and sterilized one time only, and all accessories (including gloves, face masks and goggles) were sterilized or disinfected. Test Series 2 was performed in two parts with five different test subjects in each part, i.e., a total of ten test subjects. The second part was expanded to also include comparative tests of one pharmaceutical clothing system and one surgical clothing system.

In part 1, each test subject performed 12 test sequences with new, modern cleanroom clothing systems. Included in the tests were small variations such as with and without goggles, different face masks, and different hood sizes. In part 2, each test subject performed six test sequences with new, modern cleanroom clothing systems, with variations in fabrics; and as a comparison two sequences with a pharmaceutical clothing system and a surgical clothing system, respectively. Under the cleanroom coverall was worn long-sleeved cleanroom undershirts with elastic cuffs at the wrists. In part 2 long-legged cleanroom underpants with elastic cuffs at the ankles were additionally worn. Special cleanroom socks were used in the inner shoes. All modern cleanroom clothing systems tested reflect state-of-the-art in cleanrooms and the current Good Manufacturing Practice (CGMP) of pharmaceutical production of sterile drugs. A total of 110 sequences were carried out by ten test subjects. The test subjects were all men. The data derived from Test Series 2 were analyzed using common statistical methods described in statistical textbooks including Daniel (1974), Moore (1997) and Aczel (1999).

In part 1, the test subject wore new, modern cleanroom clothing systems, long-sleeved cleanroom undershirts with elastic cuffs at the wrists, cleanroom socks, and inner shoes. Small variations in accessories such as with and without goggles, different face masks and different hood sizes were tested. From each test sequence, the mean values of the total number of airborne particles equal to and larger than 0.5 μm per cubic meter and the number of airborne aerobic CFU per cubic meter are shown in Figures 5 and 6, respectively. Grand mean values (GMV) based on results from the different test subjects are also shown in Figures 5 and 6.

Figure 5. Mean values from each test sequence and grand mean values (GMV) of total number of airborne particles ≥ 0.5 μm per cubic meter, measured with five different test subjects dressed in modern cleanroom clothing systems with small variations (with and without goggles, different face masks and different hood sizes).

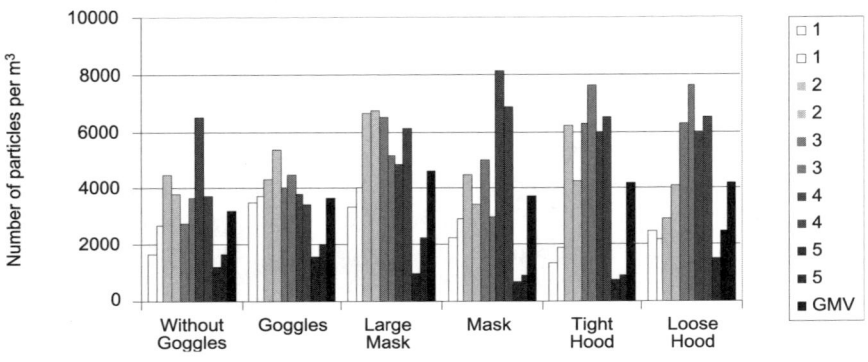

The mean values from each test sequence shown in Figures 5 and 6 respectively have been used to determine if differences in performance occur between the small variations in accessories by using the analysis of variance (ANOVA) method. With a 5% level of significance there were no significant differences between the small variations in accessories such as with and without goggles, different face masks and different hood sizes.

The second part was performed with three different types of new cleanroom clothing systems in combination with cleanroom underwear (shirt and pants) with long sleeves and long legs. The three types of cleanroom clothing systems (variations in fabrics) are called coverall Japan, coverall US and coverall Tyvek®. As a further basis for a comparison, one pharmaceutical clothing system (LIF-system) and one surgical clothing system were also tested. Figure 7 shows the mean values from each test sequence and the grand mean values of total number of airborne particles equal to and larger than 0.5 μm per cubic meter. Figure 8 shows the mean values from each test sequence and the grand mean values of number of aerobic CFU per cubic meter. The CFU data here are obtained from measurements performed with both the FH3 slit-to-agar sampler and the Andersen 6-Stage Sieve Sampler.

Figure 6. Mean values from each test sequence and the grand mean values (GMV) of airborne aerobic CFU per cubic meter during the same tests shown in Figure 5.

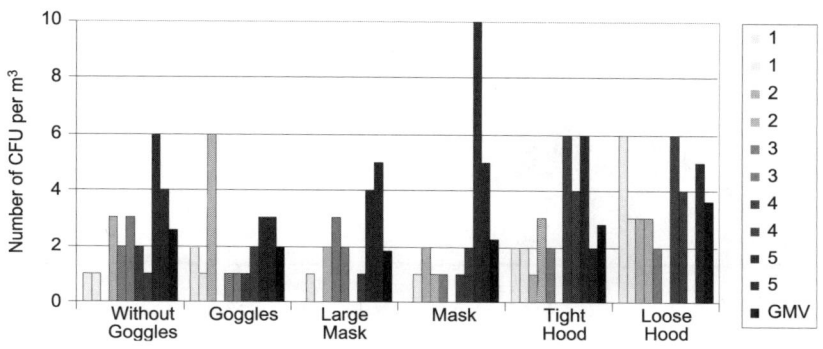

Figure 7. Mean values from each test sequence and the grand mean values (GMV) of total number of airborne particles ≥ 0.5 µm per cubic meter, measured with five different test subjects dressed with five different clothing systems.

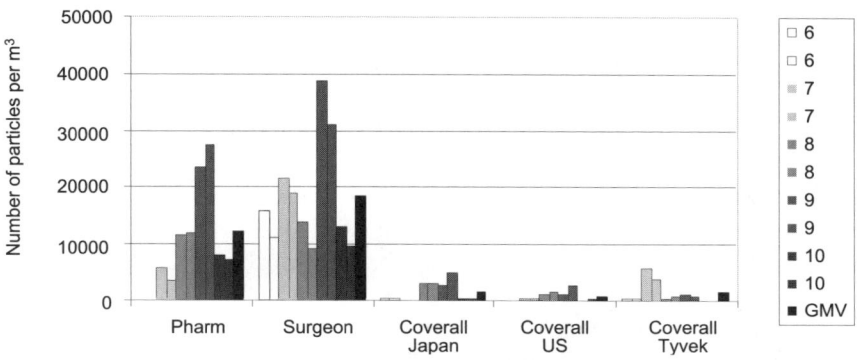

Figure 8. Mean values from each test sequence and the grand mean values (GMV) of airborne aerobic CFU per cubic meter during the same tests shown in Figure 7.

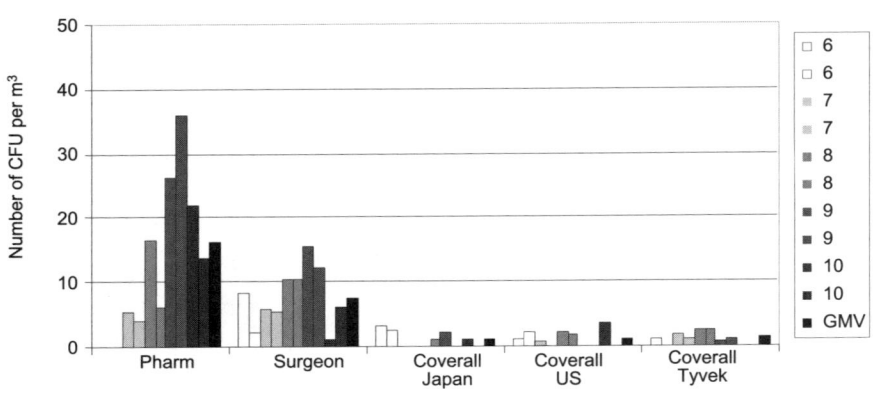

In the same manner, as performed to evaluate differences in cleanroom garment accessories, the ANOVA analysis with 5% level of significance was used to determine if there are any differences in performance among the clothing systems described in the second part. The data clearly indicate that there are significant differences between the cleanroom coverall systems and the pharmaceutical as well as surgical clothing systems. The ANOVA analysis confirms that there is a significant difference between the pharmaceutical clothing system and the surgical clothing system, but there is no significant difference among the coverall Japan, coverall US, and coverall Tyvek®, respectively.

The ratios between the total number of airborne particles ($\geq$ 0.5 µm and $\geq$ 5 µm) per cubic meter and the number of aerobic CFU per cubic meter for the clothing systems used for aseptic production of sterile products are shown in Table 5. Mean values and sample standard deviations for the experimental data are also given.

The mean value of the ratio between particles equal to and larger than 0.5 µm and airborne aerobic CFU in this instance is estimated to be about 1,500 ± 500 to 1. This value is in the lower range of the ratio given in Test Series 1 for cleanroom clothing. The lower values of the ratio in Test Series 2 are explained by the fact that only new modern cleanroom clothing systems with few washing/sterilizing cycles were used.

A ratio between particles equal to and larger than 5 µm and airborne aerobic CFU is more difficult to estimate. This might be because larger particles show different physical properties from smaller particles. The ratio between particles equal to and larger than 5 µm and airborne aerobic CFU varies according to values given for cleanroom clothing in Test Series 1 in the range 10 to 1 and 190 to 1. The values in Test Series 2 are in the lower part of the above-given range and could also here be explained by the fact that the clothing systems in Test Series 2 are new modern cleanroom clothing systems.

Table 5. Ratios between the total number of airborne particles (≥ 0.5 μm, ≥ 5 μm) per cubic meter and the number of aerobic CFU per cubic meter for the clothing systems used for aseptic production of sterile products.

Clothing system	Ratio of No of particles (≥ 0.5 μm) No. of CFU	Ratio of No of particles (≥ 5 μm) No. of CFU
Cleanroom clothing without goggles	1.26×10^3	18.4
Cleanroom clothing with goggles	1.81×10^3	27.7
Cleanroom clothing with large face mask	2.59×10^3	32.4
Cleanroom clothing with ordinary face mask	1.64×10^3	20.5
Cleanroom clothing with tight hood	1.49×10^3	18.0
Cleanroom clothing with loose hood	1.18×10^3	14.1
Cleanroom clothing coverall Japan	1.37×10^3	31.0
Cleanroom clothing coverall US	0.72×10^3	14.5
Cleanroom clothing coverall Tyvek ®	1.31×10^3	24.6
Mean value with standard deviation	$(1.49 \pm 0.52) \times 10^3$	22.4 ± 6.9

SOURCE STRENGTH – NEW CLOTHING SYSTEMS

It is important in pharmaceutical production to have knowledge about the strength of the contamination source, people. The measurements described in Test Series 2 have been performed with ten different test subjects using modern cleanroom clothing of the kind commonly used in facilities for aseptic production of sterile products. By using the air volume flow (0.22 m^3/s) in the dispersal chamber and the measured concentrations, the source strengths of each test subject can be estimated. Figures 9, 10 and 11 show the source strength of airborne particles equal to and larger than 0.5 μm per second, particles equal to and larger than 5 μm per second and aerobic CFU per second, respectively.

Figure 9. The source strengths of airborne particles ≥ 0.5 μm per second and the calculated mean value (MV) for ten test subjects dressed in modern cleanroom clothing systems.

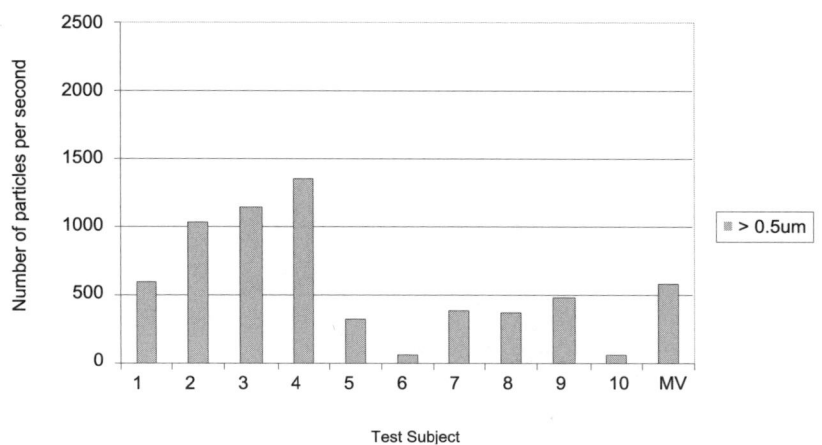

Figure 10. The source strengths of airborne particles ≥ 5 µm per second and the calculated mean value (MV) for ten test subjects dressed in modern cleanroom clothing systems.

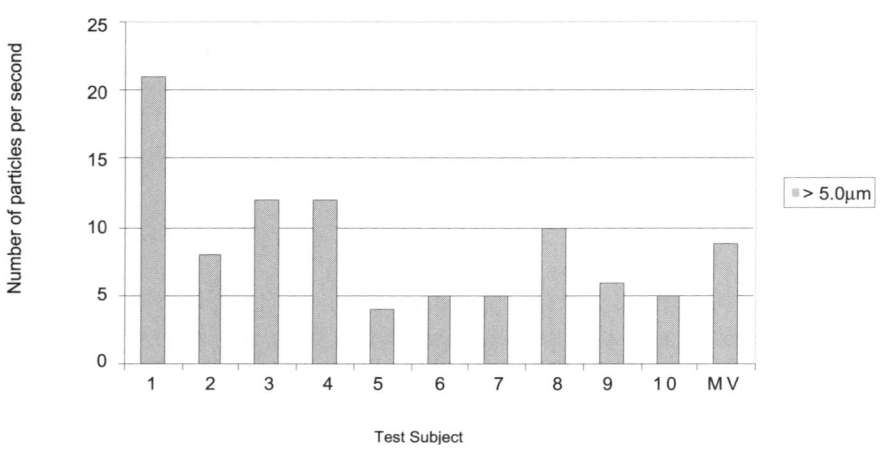

Figure 11. The source strengths of aerobic airborne CFU per second and the calculated mean value (MV) for ten test subjects dressed in modern cleanroom clothing systems.

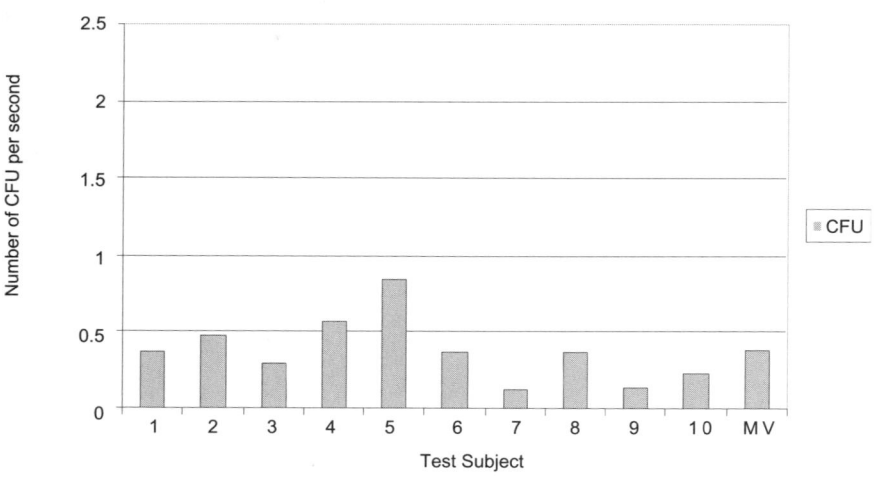

Table 6. Mean values and sample standard deviations of the source strength from ten test subjects dressed in modern cleanroom clothing.

Airborne particles and CFU	Number per second	
	Mean value	Standard deviation
Particles ≥ 0.5 µm	585	449
Particles ≥ 5 µm	9	5
Aerobic CFU	0.38	0.22

The mean values and sample standard deviations from the values given in Figures 9, 10 and 11 are summarized in Table 6. As previously mentioned, the airborne CFU have in Test Series 2 been measured with both an FH3 slit-sampler and an Andersen 6-Stage Sampler. The viable particle size distribution, according to the Andersen 6-Stage Sampler, calculated as percentage of airborne aerobic CFU found with the Andersen 6-stage Sampler is shown in Figure 12. This figure shows that more than half the viable aerobic particles recovered are smaller than 4.7 µm according to the size distribution from Andersen 6-Stage Air Sampler. This is in agreement with results from a daycare center presented by Reinmüller (2001). This implies that the total number of particles equal to and larger than 0.5 µm offers a better overview of the total situation than that of particles equal to and larger than 5 µm. In the following, only particles equal to and larger than 0.5 µm will be considered.

Figure 12. Viable particle size distribution in percent of airborne aerobic CFU, measured with an Andersen 6-stage sampler, during tests with new modern cleanroom clothing systems.

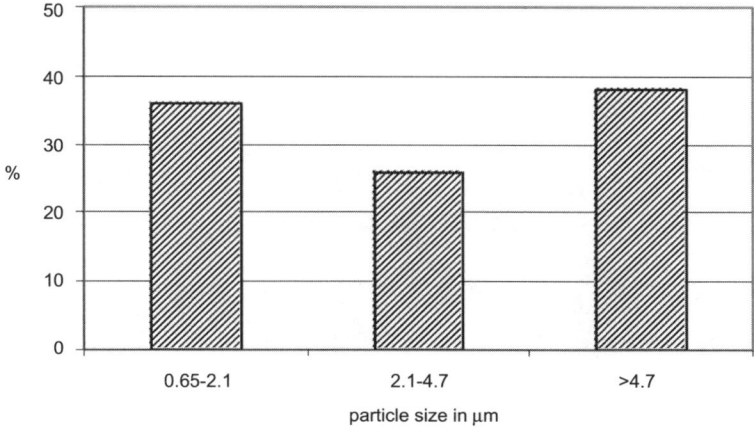

Figures 9, 10 and 11 show that the source strengths of test subjects 1–5 are generally higher than that of test subjects 6–10. These differences are statistically significant based upon ANOVA analysis with 5% level of significance. A possible explanation for these findings is that test subjects 6–10 performed the tests in part 2 wearing additional long-legged cleanroom underpants.

Furthermore, the ANOVA analysis at a 5% level of significance shows that there are significant differences among the ten test subjects (data from Figs 9 and 11). The variations observed are much larger than the variations found among the nine different cleanroom clothing systems (data from Figs 5, 6, 7 and 8). The larger variations among the test subjects, compared to the cleanroom clothing systems, would justify the treatment of data from part 1 (test subjects 1–5) and part 2 (test subjects 6–10) as one experimental group, when predicting source strength.

Consequently, t-distribution calculations of confidence intervals for a population mean of the source strengths have been performed in three steps. The first step is calculations for test subjects 1–5 (part 1) and test subjects 6–10 (part 2) and all ten test subjects (part 1 and 2). The results of these calculations at a confidence interval of 95% are shown in Table 7 for particles equal to and larger than 0.5 μm per second and in Table 8 for aerobic airborne CFU per second.

Table 7. The 95% confidence intervals (t-distribution) for source strengths of particles $\geq$ 0.5 μm per second from test subjects dressed in new modern cleanroom clothing systems. Part 1 consists of five test subjects without, and part 2 of five test subjects with long-legged cleanroom underpants, and part 1 and 2 consists of all ten test subjects.

Test Series 2	Particle ($\geq$ 0.5 μm) per second		
Part (No. of test subjects)	Mean value	Confidence interval	
		Lower	Upper
Part 1 (n=5)	892	367	1,417
Part 2 (n=5)	278	31	526
Part 1 and 2 (n=10)	585	264	907

It should be noted that the mean values for the experimental groups given in Tables 6, 7 and 8 are based upon individual mean values from five and ten test subjects, respectively. The individual mean value from each test subject is based upon results from test sequences of the tested cleanroom clothing systems.

On the other hand, the data in Table 5 are based on mean values from nine tested variations of cleanroom clothing systems, where each mean value is based on results from five test subjects.

Table 8. The 95% confidence intervals (*t*-distribution) for source strengths of airborne aerobic CFU per second from test subjects dressed in new modern cleanroom clothing systems. Part 1 consists of five test subjects without, and part 2 of five test subjects with long-legged cleanroom underpants, and part 1 and 2 consists of all ten test subjects.

Test Series 2	Aerobic CFU per second		
Part (No. of test subjects)	Mean value	Confidence interval	
		Lower	Upper
Part 1 (n=5)	0.51	0.24	0.78
Part 2 (n=5)	0.24	0.10	0.39
Part 1 and 2 (n=10)	0.38	0.22	0.53

Data in Table 5 give the *t*-distribution estimation of the 95% confidence interval for the ratio between the total number of particles equal to and larger than 0.5 µm and number of aerobic airborne CFU, which becomes 1,490 and 400 to 1. This implies that there is 95% confidence that the population mean ratio is between 1,090 and 1,890 to 1. These two values, and the values of confidence intervals from Table 7 and Table 8, are shown in a graphic form in Figure 13.

In Figure 13 it can be seen that the three mean values are within the area between the lines of the upper and the lower level of the ratio between airborne particles equal to and larger than 0.5 µm and airborne aerobic CFU. The ranges of the 95% confidence intervals plotted for source strengths depending on individuals are wider than that of the plotted ratio based on various cleanroom clothing systems.

If a general situation could be identified, mean values from Figure 13 might be used. On the other hand, when assessing particular risk situations it might be advisable to overestimate the risk by using the maximum values of both particles and CFUs (right upper corner values). Results given in the literature (for example, Gustavsson, 1999; Whyte, 1999; Tammelin et al., 2000) have been attained in different ways. This means that values of the source strength people dressed in a clothing system often have to be estimated from theoretical calculations; and values from such estimations should only be considered as indicative.

A comparison made from published data and from results calculated from data in Figures 2, 3, 7 and 8 and Tables 3 and 6 (previously reported by Reinmüller, 2001) are shown in Table 9.

Figure 13. The 95% confidence intervals (*t*-distribution) for source strengths of total airborne particles ≥ 0.5 µm per second and airborne aerobic CFU per second.

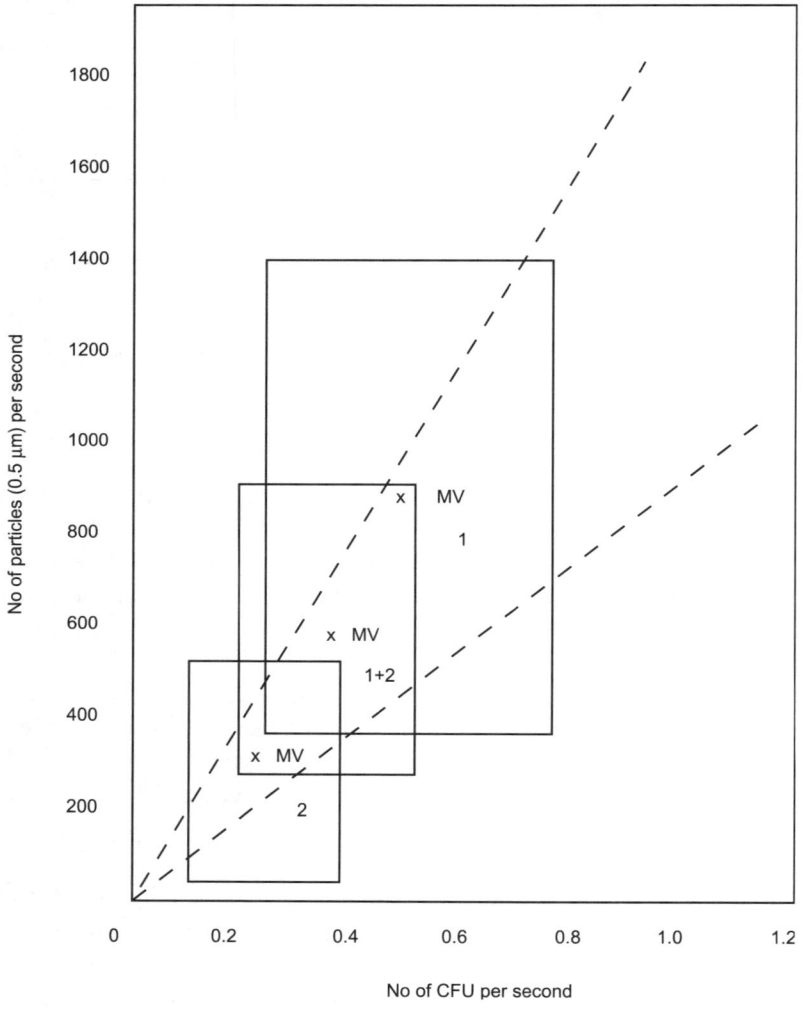

Note: Rectangles; data from Tables 7 and 8, part 1, part 2, and part 1 and 2. The lower level and the upper level of the 95% confidence interval (*t*-distribution) for the ratio between airborne particles ≥ 0.5 µm and airborne aerobic CFU, (dashed lines; data from Tab. 5).

The data in Table 9 show that values concerning "lab coat – disp coat" are in a similar range for all four tests. Values of aerobic CFUs from tests with surgical clothing are in comparable level. The particle data from modern cleanroom clothing systems (reported here) are in a lower level than those reported by Gustavsson (1999) and Whyte (1999), while reported aerobic CFU data are in the same range. It is possible that the differences shown are dependent upon differences in quality of the clothing systems as well as differences in the measuring devices and test conditions.

Table 9. Comparison of data of the source strength (particles per second and CFU per second) people dressed in various clothing systems.

| Clothing system | Contaminant | Particles per second and CFU per second | | | |
		Gustavsson (1999)	Whyte (1999)	Tammelin et al. (2000)	Data from Reinmüller (2001)
Lab coat Disp. coat Conv. scrub	Part ≥ 0.5 µm	10,000 – 40,000	33,000	–	26,700
	Part ≥ 5 µm	–	5,000	–	1,780
	CFU	–	2.7	1.88 *	4.6
Tightly woven scrub Surgical system	Part ≥ 0.5 µm	–	–	–	4,060
	Part ≥ 5 µm	–	–	–	270
	CFU	–	–	0.77 *	1.7
High quality cleanroom clothing system	Part ≥ 0.5 µm	1,000 – 10,000	16,000 *	–	585
	Part ≥ 5 µm	–	600 *	–	9
	CFU	–	0.22 *	–	0.38

*   Theoretical calculated data.
Whyte (1999) estimated from penetration tests.
Tammelin et al. (2000) estimated by using given values of air flows and assuming dilution principle.

TEST SERIES 3

In Test Series 3 the standardized cycles of movements have been performed in the same manner as described in Test Series 2. The tests were performed with two different types of modern cleanroom clothing systems in combination with two variations of underwear.

The first variation consisted of a long-sleeved cleanroom undershirt with elastic cuffs at the wrists and short-legged underpants. The second consisted of a long-sleeved cleanroom undershirt with elastic cuffs at the wrists and long-legged cleanroom pants with elastic cuffs at the ankles. The two types of cleanroom clothing coveralls (variations in treatment, see Appendix) are identified as coverall Japan and coverall US.

As a comparison, tests were also performed with one surgical clothing system. This clothing system was tested after having been washed 25 times and 50 times, respectively.

The cleanroom clothing systems were tested after having been washed and sterilized 25 times and 50 times, respectively. All accessories – such as head covers, disposable gloves face masks – were sterilized and the shoes disinfected.

The washing and sterilizing cycle used here for cleanroom clothing consists of the washing process at a temperature of 73 ± 2 °C for 10 minutes, rinsing steps, and a dry-tumbling with HEPA-filtered warm air. The sterilization cycle is carried out in an autoclave with steam at 121 °C for 20 minutes.

The cleanroom coverall weaves are both densely woven continuous filament polyester fabric with electro-conductive fibers (ECF) for discharge of static electricity. One of the weaves is fluid repellent and contains a durable antimicrobial treatment. The underwear weave is densely woven continuous filament 100% polyester fabric containing ECF and treated with a durable antimicrobial finish. The surgical clothing material is a mix of 50% cotton and 50% polyester.

HelmkeDrum tests and BubblePoint tests have also been performed after 25 and 50 washing/sterilizing cycles, respectively. HelmkeDrum tests have been carried out by Berendsen in Nyköping, Sweden, and BubblePoint tests by Fristads AB in Fristad, Sweden.

Figures 14–16 illustrate results obtained from clothing systems after 25 washing/sterilizing cycles. As mentioned earlier, the surgical clothing system is not sterilized after washing. In the figures, coverall Japan and coverall US are identified as J and US, respectively.

Figure 14. Number of airborne total particles ≥ 0.5 μm per cubic meter measured from tests in the body box with five different test subjects with clothing system washed 25 times.

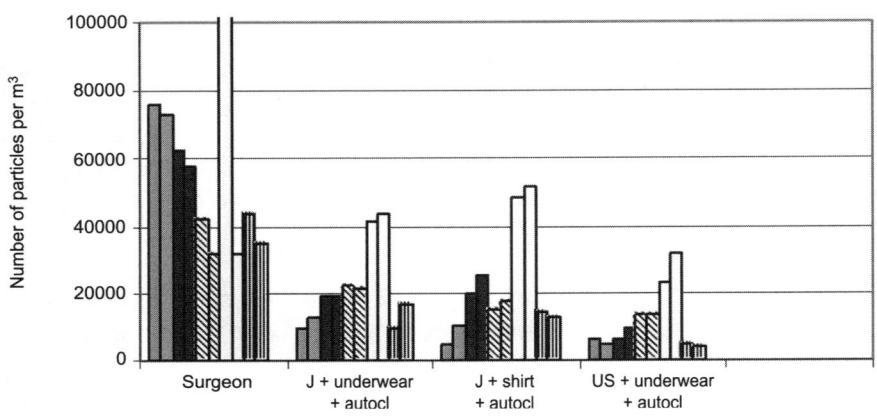

Results from 50 washing/sterilizing cycles are shown in Figures 17–19. These also show the results from a comparison of one cleanroom clothing system (Japan coverall in combination with cleanroom underwear) washed 50 times but not sterilized (autoclaved) in similarity with the surgical clothing system.

Figure 15. Number of airborne total particles ≥ 5 µm per cubic meter measured from tests in the body box with five different test subjects with clothing system washed 25 times.

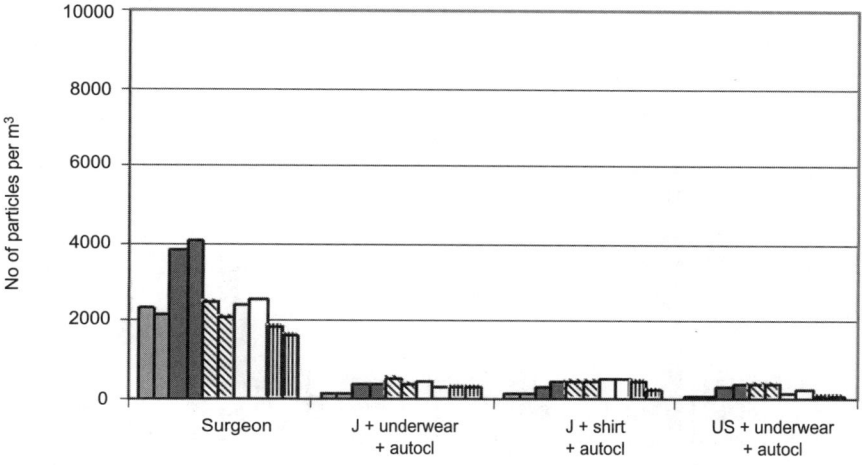

Figure 16. Number of airborne CFU per cubic meter measured from tests in the body box with five different test subjects with clothing system washed 25 times.

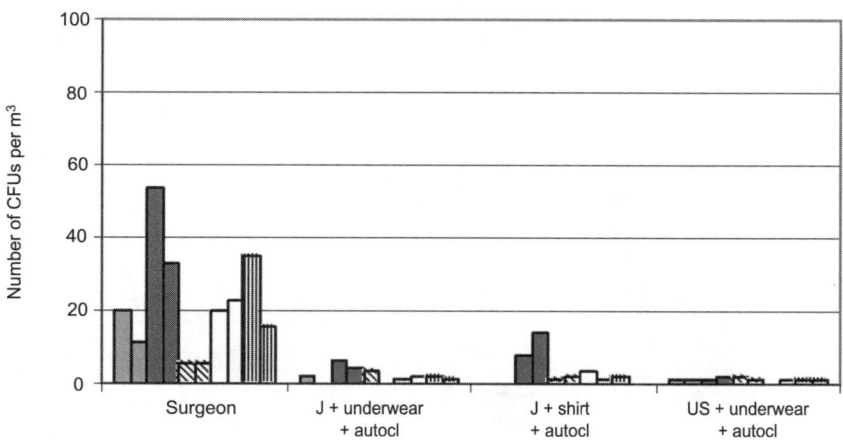

Cleanroom Clothing Systems

Figure 17. Number of airborne total particles ≥ 0.5 µm per cubic meter measured from tests in the body box with five different test subjects with clothing system washed 50 times.

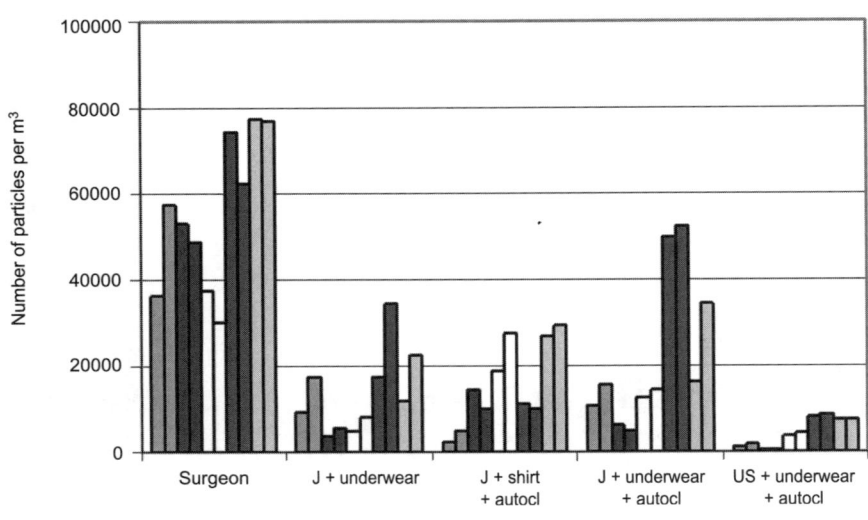

Figure 18. Number of airborne total particles ≥ 5 µm per cubic meter measured from tests in the body box with five different test subjects with clothing system washed 50 times.

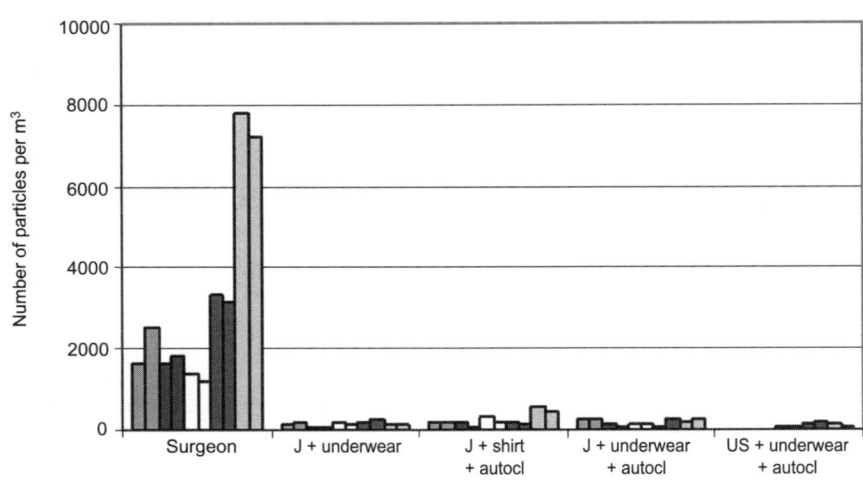

Figure 19. Number of airborne CFU per cubic meter measured from tests in the body box with five different test subjects with clothing system washed 50 times.

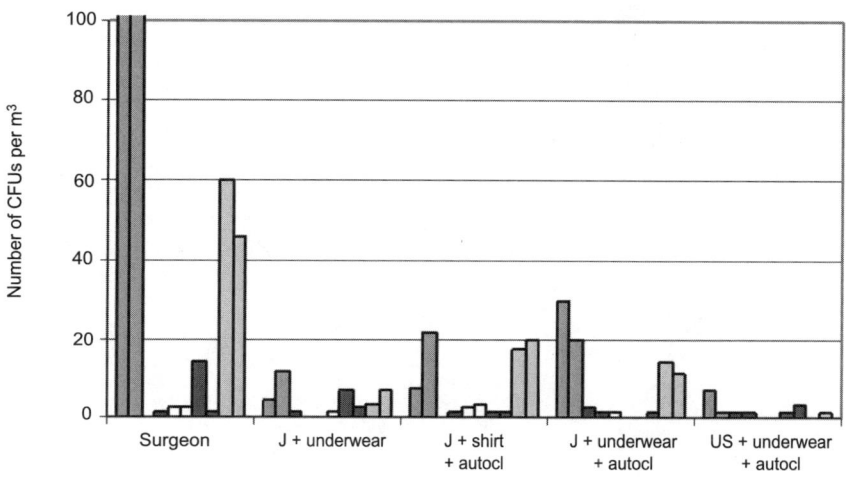

The data from Figures 14–19 clearly indicate that there are significant differences between the cleanroom clothing systems and the surgical clothing system. In comparison with the cleanroom clothing systems the data indicate that the US coverall in combination with cleanroom underwear gives the best results.

After 50 washing cycles the Japan coverall in combination with cleanroom underwear shows lower protection efficiency when washed and sterilized (autoclaved) than when only washed.

The strength of the contamination source, people – here called source strength – can be estimated by using the air volume flow (0.22 m^3/s) in the dispersal chamber and the measured concentrations. The tests with modern cleanroom clothing systems have been performed with five different test subjects each for 25 and 50 washes. Figures 20–25 show the test subjects' source strength during the tests with cleanroom clothing systems washed 25 and 50 times, respectively.

Figure 20. The source strength of airborne particles ≥ 0.5 μm per second and the calculated mean value (MV) for five test subjects dressed in cleanroom clothing systems washed 25 times.

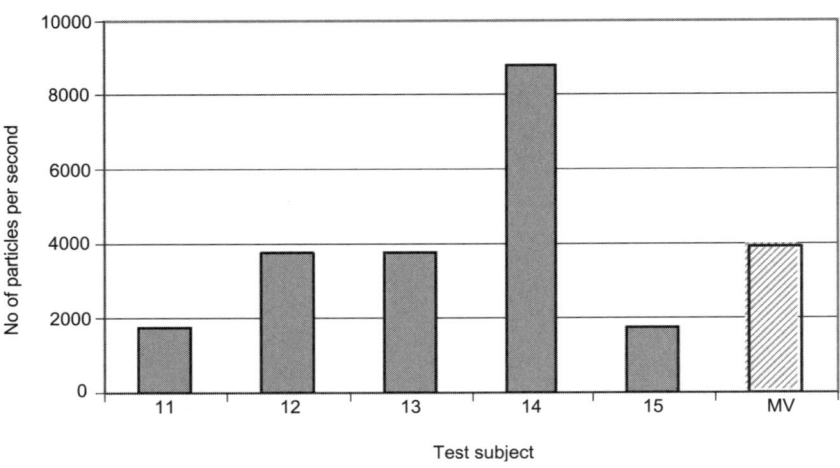

Figure 21. The source strength of airborne particles ≥ 5 μm per second and the calculated mean value (MV) for five test subjects dressed in cleanroom clothing systems washed 25 times.

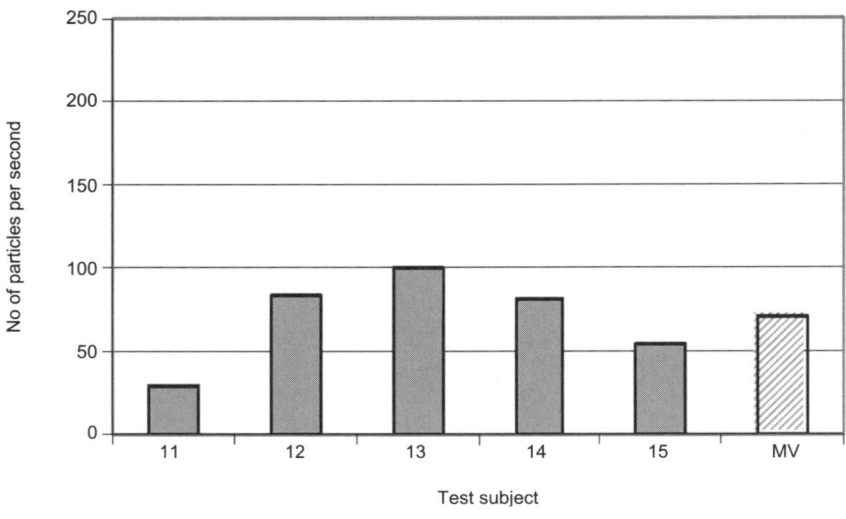

Figure 22. The source strength of airborne CFU per second and the calculated mean value (MV) for five test subjects dressed in cleanroom clothing systems washed 25 times.

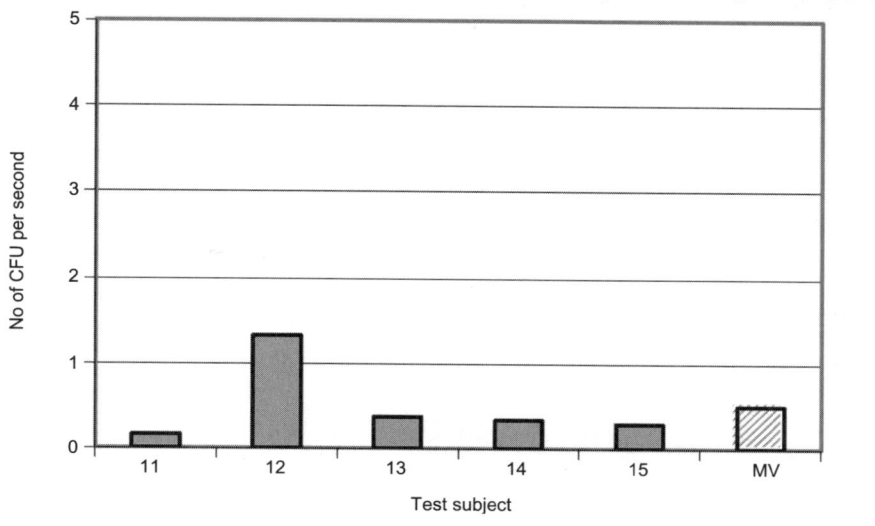

Figure 23. The source strength of airborne particles ≥ 0.5 μm per second and the calculated mean value (MV) for five test subjects dressed in cleanroom clothing systems washed 50 times.

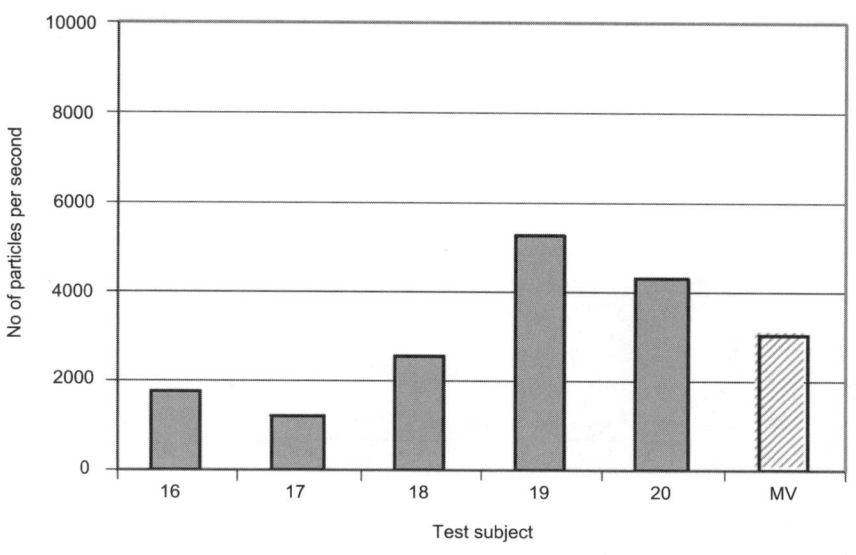

Figure 24. The source strength of airborne particles ≥ 5 μm per second and the calculated mean value (MV) for five test subjects dressed in cleanroom clothing systems washed 50 times.

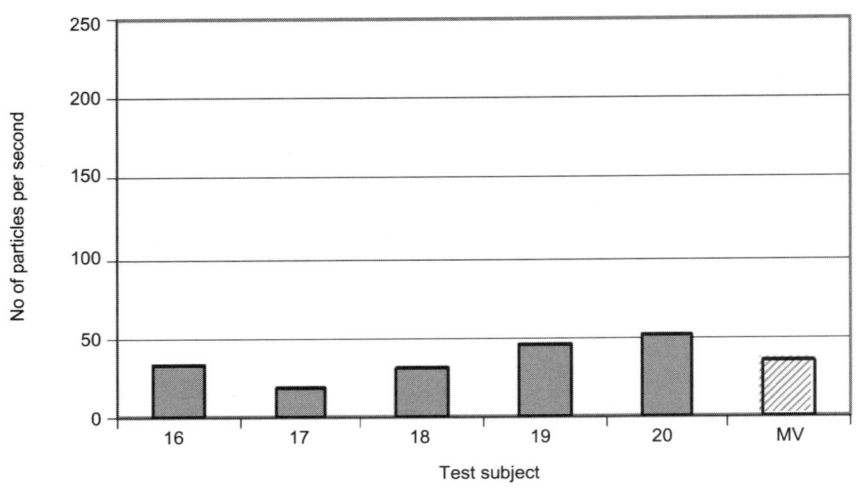

Figure 25. The source strength of airborne CFU per second and the calculated mean value (MV) for five test subjects dressed in cleanroom clothing systems washed 50 times.

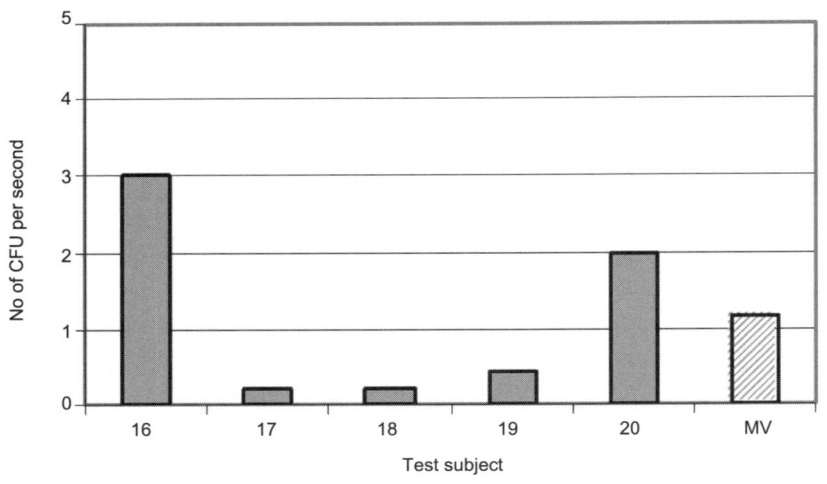

Figures 20–25 show that the differences are small between mean values of cleanroom clothing system washed 25 times and 50 times.

Figure 26. Results from HelmkeDrum tests from cleanroom clothing fabrics (coverall US and Japan) washed/sterilized once, 25 times and 50 times, respectively.

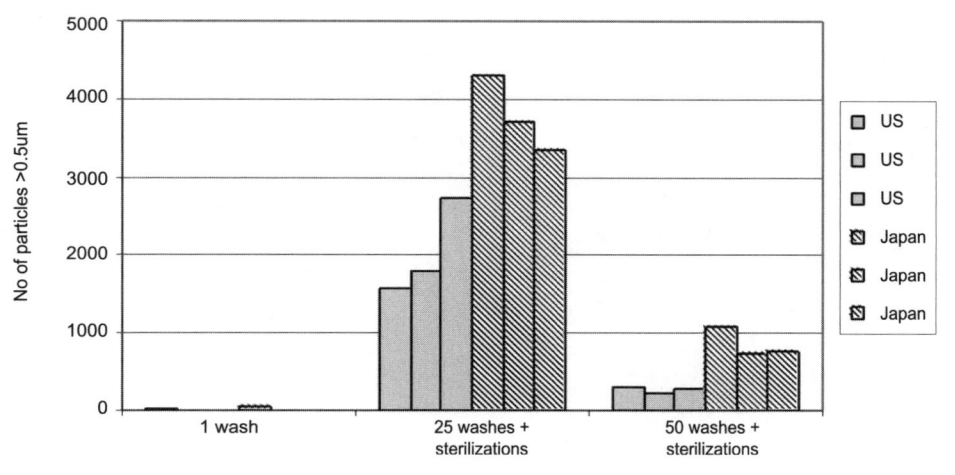

Figures 26 and 27 show results from HelmkeDrum tests and BubblePoint tests from cleanroom clothing systems washed/sterilized once, 25 times and 50 times, respectively. The results in these figures indicate that there seems to be a relation between data from BubblePoint tests and data from dispersal chamber tests, while a relation between data from HelmkeDrum tests and data from dispersal chamber tests is difficult to be found.

A comparison of source strengths made from data in Table 9 and from the discussed results is shown in Table 10.

Figure 27. Results from BubblePoint tests from cleanroom clothing fabrics (coverall US and Japan) washed/sterilized once, 25 times and 50 times, respectively.

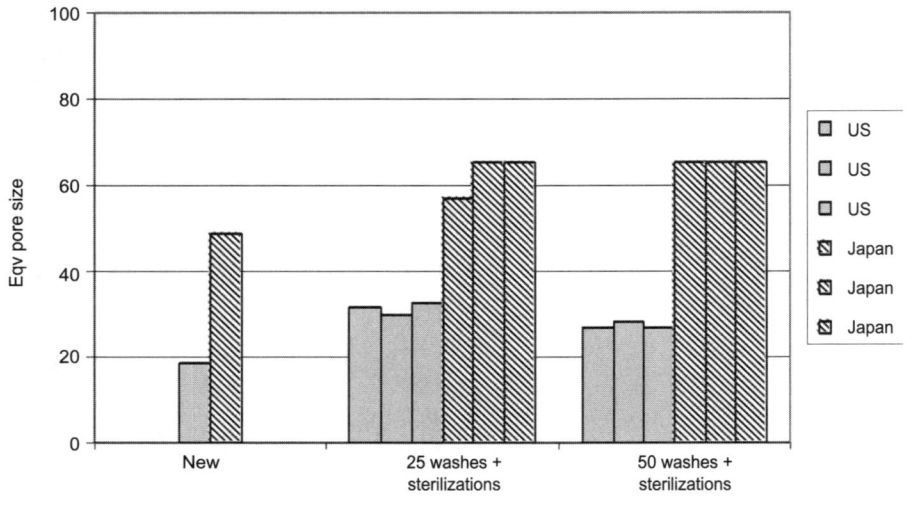

Table 10. Comparison of data (mean values) of the source strength (particles per second and CFU per second) people dressed in various clothing systems washed/sterilized once, 25 times and 50 times, respectively.

Clothing system	Contaminant	Particles per second and CFU per second		
		1 wash	25 washes	50 washes
Surgical clothing system	Particles ≥ 0.5 μm	4,060	13,875	12,207
	Particles ≥ 5 μm	270	535	698
	CFU	1.7	4.2	9.0
High quality cleanroom clothing system	Particles ≥ 0.5 μm	585	3,950	2,860
	Particles ≥ 5 μm	9	70	36
	CFU	0.38	0.49	1.14

DISCUSSION OF TEST RESULTS

Cleanroom technology of today provides tailor-made systems for most demands; therefore, unequipped rooms are generally not considered to be a problem. However, the situation may change dramatically when cleanrooms are equipped and run with machinery, operators and processes that create temperature differences.

Within the cleanroom, people are the main source of airborne microbial contamination. Values of the strength of the contamination source (people) dressed in new modern cleanroom clothing systems have been estimated from measurements performed in a dispersal chamber. These values are lower than values given in literature. Results from performed tests show no significant differences of the released airborne particulate contamination levels among small variations in accessories such as with and without goggles, different face masks and different sizes of hoods. Results from a comparison of two types of cleanroom underwear used together with the cleanroom coverall show significantly lower levels of released airborne particulate contamination when long-sleeved cleanroom undershirts were used in combination with long-legged cleanroom underpants.

Table 10 shows that for the surgical clothing system and the cleanroom clothing systems there is an increase of the source strength after 25 and 50 washing/sterilizing cycles compared with new clothing systems. However, this indicates that the studied cleanroom clothing systems in combination with cleanroom underwear could still be used at least up to 50 washes/sterilizing cycles with acceptable protection efficiency.

During this study, defects occurred in the function of zippers and snap fasteners. To expand the time of use for the cleanroom clothing systems it might be necessary to improve the quality and function of details such as zippers, snap fasteners and clasps.

Results from a comparison of two types of cleanroom underwear used together with the cleanroom coverall show lower levels of released airborne contamination when long-sleeved cleanroom undershirts were used in combination with long-legged cleanroom underpants.

The results from the BubblePoint tests, as well as the tests in the dispersal chamber, indicate that the material used in the US coverall give a higher reduction of airborne contamination than that used in the Japan coverall.

It should be mentioned that it is difficult to find, in this investigation a relationship between data from HelmkeDrum tests and data from dispersal chamber tests. In spite of this, it can be noted that the particle levels in the HelmkeDrum tests (Fig. 26) as well as in the dispersal chamber tests (Tab. 10) reach higher values at 25 washing/sterilizing cycles than after 50 washing/sterilizing cycles. This might be explained by the fact that after a certain number of washing/sterilizing cycles the fabrics release particles. With time, the released particles seem to be washed away from the fabrics.

Values presented indicate that a relationship exists between the number of particles equal to and larger than 0.5 μm per volume unit of air and the number of aerobic airborne CFU per volume of unit air. In a typical cleanroom environment, with people dressed in new modern cleanroom clothing systems (washed/sterilized once) as the main contamination source, it seems to be possible to establish a relationship at a ratio of approximately 1,500 to 1. When the cleanroom clothing systems are washed/sterilized several times this relationship will increase, where the values of the ratio is estimated to be less than 10,000 to 1.

GUIDELINES FOR PHARMACEUTICAL PRODUCTION

GENERAL

Since production of sterile drugs for global sale needs manufacturing conditions fulfilling the strengthened national and international requirements of cleanliness and documentation, production of sterile drugs today is performed in classified cleanrooms.

In ISO 14644-1 "Cleanrooms and associated controlled environments" a cleanroom is defined as:

> A room in which the concentration of airborne particles is controlled, and which is constructed and used in a manner to minimize the introduction, generation and retention of particles inside the room and in which other relevant parameters, e.g., temperature, humidity, and pressure, are controlled as necessary.

The development of cleanrooms started after the Second World War. In the 1950s to 1960s, the high efficiency particulate air filter (HEPA-filter[1]), first developed for military use, found applications within operating room ventilation systems and in pharmaceutical production areas for sterile drugs. Pharmaceutical industry cleanrooms can be of different types but they all have increased air supply compared to ordinary rooms (such as offices or schools), terminal HEPA filters and room pressurization.

[1.] HEPA-filter "A through-away extended media dry type filter in a rigid frame having a minimum particle efficiency of 99.97% for 0.3 μm thermally generated DOP-particles or specified aerosol, and a maximum pressure drop of 2.54 water gauge, when tested at rated air flow capacity" (IES-RP-CC-001-3, 1992).

In the early 1960s, the unidirectional or "laminar flow" concept was realized. Unidirectional airflow[2] (uniform velocity and direction) is used when high degrees of cleanliness are required. The expression "laminar flow" in pharmaceutical manufacturing is sometimes used in place of the term unidirectional flow. Unidirectional flow units (UDF units) are used to increase the cleanliness in smaller, critical regions. A further development is the isolator, which has rigid or flexible wall systems and transfer and manipulation systems applied to specific processes and is supplied with HEPA-filtered air.

With the supply air filtered through HEPA-filters, within the cleanroom the main sources of airborne particles are people and operating machinery. People are the principal source of airborne viable particles. In order to protect the air quality, people should wear specially designed cleanroom clothing systems. The prime function of cleanroom clothing is to act as a body filter and be designed in such a way to prevent unfiltered body emission to be dispersed in the cleanroom. In the 1960s, cleanroom clothing was often woven cotton or woven cotton-polyester clothing. Today cleanroom clothing used in cleanrooms can be either disposable or reusable. The disposable or limited-use clothing is usually made from a non-woven material. The reusable clothing is processed (washed and sterilized) regularly and usually made from tightly woven synthetic fabrics. Natural fabrics made from e.g., cotton, are no longer used in pharmaceutical cleanrooms. Present cleanroom clothing, with high filtration efficacy and often used in combination with cleanroom underwear, has been shown to reduce both the number of total particulates (≥ 0.5 μm) and the number of viable particles (CFUs) emitted by people in the cleanroom.

As the technical solutions have improved, the GMP requirements regarding manufacturing conditions (airborne particulates; total number and viable number) of sterile drugs have increased. However, there is not yet a harmonization between different GMP requirements.

For pharmaceutical cleanrooms, the most used guidelines and requirements are:

FDA CGMP   Applies to products sold in the USA

EU GMP Annex 1   Applies to products sold in the EU region

---

[2]. Unidirectional airflow "Rectified airflow through the entire section of a clean zone with a steady velocity and approximately parallel streamlines. This type of flow results in a direct transport of particles from the clean zone." ISO 14644-4.

Annex 1 of the EC Guide to Good Manufacturing Practice (GMP) provides supplementary guidance on the application of the principles and guidance of GMP to sterile products. The guidance includes recommendations on standards of environmental cleanliness for cleanrooms.

According to the Annex 1, for the manufacture of sterile medicinal products, four grades can be distinguished:

Grade A: The local zone for high risk operations, e.g., filling zone, stopper bowls, open ampoules and vials, making aseptic connections. Normally such conditions are provided by a laminar air flow work station. Laminar air flow systems should provide a homogeneous air speed in a range of 0.36–0.54 m/s (guidance value) at the working position in open clean room applications.

The maintenance of laminarity should be demonstrated and validated.

A unidirectional air flow and lower velocities may be used in closed isolators and glove boxes.

Grade B: For aseptic preparation and filling, this is the background environment for the grade A zone.

Grades C and D: Clean areas for carrying out less critical stages in the manufacture of sterile products.

In Tables 11, 12 and 13 the maximum permitted or recommended number of total particles and viable particles are compared, respectively, valid December 2003. However, in the regulatory requirements it can be noted that airborne particles (≥ 5 μm) are only considered in the EU GMP Annex 1.

Neither ISO 14644-1 (1999) nor 14698-1 (2003) defines number of airborne viable particles in classified cleanrooms and associated controlled environments.

Table 11. Maximum permitted airborne particles (≥ 0.5 µm) per cubic meter in pharmaceutical cleanrooms.

	EU GMP Annex 1		US FDA and USP		ISO 14644–1	
Grade	At rest condition	Operational condition	Area	Operational condition	Class	Operational condition
Grade A	3,500	3,500	Critical	3,500	Class 5	3,520
Grade B	3,500	350,000	Background to critical area	350,000	Class 7	352,000
Grade C	350,000	3,500,000	Controlled	3,500,000	Class 8	3,520,000
Grade D	3,500,000	Not defined	—	—	—	—

Table 12. Maximum permitted airborne particles (≥ 5 µm) per cubic meter in pharmaceutical cleanrooms.

	EU GMP Annex 1		ISO 14644–1	
Grade	At rest condition	Operational condition	Class	Operational condition
Grade A	1	1	Class 5	29
Grade B	1	2,000	Class 7	2,930
Grade C	2,000	20,000	Class 8	29,300
Grade D	20,000	Not defined	–	–

To fulfil the requirements of air quality during manufacturing conditions, the outside airborne particulate contaminants are prevented from entering by use of suitably high efficiency filters (HEPA or ULPA filters). To dilute and transport out contaminants generated within the cleanroom, the number of air changes per hour should be sufficiently high. Furthermore, the cleanroom is pressurized to prevent air from less clean areas from entering. To minimize the transfer of airborne contamination from less clean areas, the entrances and exits to the cleanroom or to a suite of cleanrooms are through ventilated airlocks or changing rooms.

Table 13. Maximum permitted/recommended airborne viable particles (CFU) per cubic meter in pharmaceutical cleanrooms.

EU GMP Annex 1		FDA 1987 and Draft 2002		USP 27 NF-22	
Grade	Operational condition	Area	Operational Condition	Class US Customary	Operational Condition
Grade A	<1	Critical	<3 (1)*	100	<3
		Background to critical area Class US Customary 1,000	7		
Grade B	10	Background to critical area Class US Customary 10,000	18 (10)*	10,000	20
Grade C	100	Controlled area Class US Customary 100,000	88 (100)*	100,000	100
Grade D	200	–	–	–	–

* recommended by PQRI, February 2003 and included in FDA Draft 2003.

Contamination sources and suitable actions to prevent contamination within the pharmaceutical cleanroom are as follows:

Air Should be filtered and the filters integrity tested at appropriate intervals (including e.g., air, gases and pressurized air).

Water Should be filtered and the filters integrity tested at appropriate intervals.

Material Should be cleaned, sterilized or disinfected.

Process Should be controlled with regard to generation of particles during operation.

People Should wear clean, sterile, high quality cleanroom clothing system. Entrance and exit procedures should be in writing.

Particle generating process equipment should be avoided or limited within the cleanroom; when necessary specially designed exhaust should be installed in close vicinity to the particle generation. To prevent generation of particles within the cleanroom, no particle shedding or linting material should be used inside the cleanroom.

To avoid contamination related to the process, all material (ingredients and products) transferred into the clean aseptic areas have to be sterile-filtered, or autoclaved, or in other ways cleaned and sterilized. Personnel should enter the cleanroom through appropriate changing rooms. The cleanroom clothing system has to be of high quality, clean and sterile.

Bradley et al. (1991) showed that the level of airborne micro-organisms in the filling environment has a profound effect on the level of product contamination. In the investigation of filling equipment – a blow fill seal machine – a direct relationship was reported between the extent of product contamination and the level of airborne microorganisms. When HEPA-filtered air was supplied over the filling zone a ten-fold reduction in product contamination was observed as opposed to no supply of HEPA-filtered air.

PERSONNEL AND CLEANROOM CLOTHING: EU GMP

EU GMP states in the latest revision of Annex 1 (2003) "General" part 3 that:

> Clean areas for the manufacture of sterile products are classified according to the required characteristics of the environment. Each manufacturing operation requires an appropriate environmental cleanliness level in the operational state in order to minimise the risks of particulate or microbial contamination of the product or materials being handled.
>
> In order to meet "in operation" conditions these areas should be designed to reach certain specified air-cleanliness levels in the "at rest" occupancy state. The "at rest" state is the condition where the installation is installed and operating, complete with production equipment but with no operating personnel present. The "in operation" state is the condition where the installation is functioning in the defined operation mode with the specified number of personnel working.
>
> The "in operating" and "at rest" states should be defined for each clean room or suite of clean rooms.

In notes to the classification tables the following guidance is provided:

c)   In order to reach the B, C and D air grades, the number of air changes should be related to the size of the room and the equipment and the personnel present in the room. The air system should be provided with appropriate terminal filters such as HEPA for grades A, B and C.

e)   These areas (grade A and B) are expected to be completely free from particles of size greater than or equal to 5 μm. As it is impossible to demonstrate the absence of particles with any statistical significance the limits are set to 1 particle/m^3. During the clean room qualification it should be shown that the areas can be maintained within the defined limits.

Examples of operations to be carried out in various grades are given below:

Grade	Examples of operations for terminally sterilised products
A	Filling of products, when unusually at risk
C	Preparation of solutions, when unusually at risk. Filling of products
D	Preparation of solutions and components for subsequent filling

Grade	Examples of operations for aseptic preparations
A	Aseptic preparation and filling
C	Preparation of solutions to be filtered
D	Handling of components after washing

7.   Regarding isolator technology, the air classification required for the background environment depends on the design of the isolator and its application. It should be controlled and for aseptic processing it should be at least grade D.

10.   Blow/fill/seal equipment used for aseptic production which is fitted with an effective grade A air shower may be installed in at least grade C environment, provided that grade A/B clothing is used. The environment should comply with the viable and non-viable limits at rest and the viable limit only when in operation.

Blow/fill/seal equipment used for the production of products which are terminally sterilised should be installed in at least a grade D environment.

11. Filling of products for terminal sterilisation should be carried out in at least a grade C environment.

Where the product is at unusual risk of contamination from the environment, for example because the filling operation is slow or the containers are wide-necked or are necessarily exposed for more than a few seconds before sealing, the filling should be done in a grade A zone with at least a grade C background.

12. Components after washing should be handled in at least grade D environment. Handling of sterile starting materials and components, unless subjected to sterilisation or filtration through a microorganism-retaining filter later in the process, should be done in a grade A environment with grade B background.

Preparation of solutions which are to be sterile filtered during the process should be done in a grade C environment; if not filtered, the preparation of materials and products should be done in a grade A environment with grade B background.

Handling and filling of aseptically prepared products should be done in a grade A environment with a grade B background.

Prior to the completion of stoppering, transfer of partially closed containers, as used in freeze drying should be done in a grade A environment with a grade B background.

Preparation and filling of sterile ointments, creams, suspensions and emulsions should be done in a grade A environment with a grade B background, when the product is exposed and is not subsequently filtered.

The section "Personnel" part 19–21 states that:

19. The clothing and its quality should be appropriate for the process and the grade of the working area. It should be worn in such a way as to protect the product from contamination.

The description of clothing required for each grade is given below:

Grade D: Hair and where relevant, beard should be covered. A general protective suit and appropriate shoes or overshoes should be worn. Appropriate measures should be taken to avoid any contamination coming from outside the clean area.

Grade C: Hair and where relevant, beard and moustache should be covered. A single or two-piece trouser suit, gathered at the wrists and with high neck and appropriate shoes or overshoes should be worn. They should shed virtually no fibres or particulate matters.

Grade A/B: Headgear should totally enclose hair, and where relevant, beard and moustache; it should be tucked into the neck of the suit; a face mask should be worn to prevent the shedding of droplets. Appropriate sterilised, non-powdered rubber or plastic gloves and sterilized or disinfected footwear should be worn. Trouser-legs should be tucked inside the footwear and garment sleeves into the gloves. The protective clothing should shed virtually no fibres or particulate matters and retain particles shed by the body.

20. Outdoor clothing should not be brought into changing rooms leading to grade B and C rooms. For every worker in a grade A/B area, clean sterile (sterilised or adequately sanitised) protective garments should be provided at each work session. Gloves should be regularly disinfected during operations. Masks and gloves changed at least for every working session.

21. Clean area clothing should be cleaned and handled in such a way that it does not gather additional contaminants which can later be shed. These operations should follow written procedures. Separate laundry facilities for such clothing are desirable. Inappropriate treatment of clothing will damage fibres and may increase the risk of shedding particles.

The section "Premises", part 27, states that:

27. Changing room should be designed as airlocks and used to provide physical separation of the different stages of changing and so minimise microbial and particulate contamination of protective clothing. They should be flushed effectively with filtered air. The final stage of the changing room should, in the at rest state, be the

same grade as the area into which it leads. The use of separate changing rooms for entering and leaving clean areas is sometimes desirable. In general, hand washing facilities should be provided only in the first stage of the changing rooms.

PERSONNEL AND CLEANROOM CLOTHING: FDA GMP

In FDA (1996) section 211.28, "Personnel Responsibilities", states that:

(a) Personnel engaged in the manufacture, processing, packing or holding of a drug product shall wear clean clothing appropriate for the duties they perform. Protective apparel, such as head, face, hand and arm coverings, shall be worn as necessary to protect drug products from contamination.

(b) Personnel shall practice good sanitation and health habits.

(c) Only personnel authorized by supervisory personnel shall enter those areas of the building and facilities designated as limited-access areas.

(d) Any person shown at any time (either by medical examination or supervisory observation) to have an apparent illness or open lesions that may adversely affect the safety or quality of drug products shall be excluded from direct contact with components, drug product containers, closures, in-process materials and drug products until the condition is corrected or determined by competent medical personnel not to jeopardize the safety or quality of drug products. All personnel shall be instructed to report to supervisory personnel any health condition that may have an adverse effect on drug products.

The FDA's "Guideline on Sterile Drug Products Produced by Aseptic Processing" (1987) mentions in the introduction that:

It should be noted that this document does not address several other important aspects of aseptic processing – such as employee hygiene, aseptic gowning, and clean room design.

FDA's Draft Sterile Drug Products Produced by Aseptic Processing (2002) says in the scope that it updates relative to the 1987 document and includes guidance on: personnel qualification, cleanroom classifications under dynamic conditions, room design, quality control, environmental monitoring, and review of production records (also mentioned in the FDA Draft 2003). In the section "Manufacturing Personnel" it reads:

> Personnel who have been qualified and permitted access to the aseptic processing area should be appropriately gowned. An aseptic processing area gown should provide a barrier between the body and the exposed sterilized materials, and prevent contamination from particles generated by, and microorganisms shed from, the body. Gowns need to be sterile and non-shedding, and should cover the skin and the hair. Face masks, hoods, beard/moustache covers, protective goggles, elastic gloves, cleanroom boots, and shoe overcovers are examples of common elements of gowns. An adequate barrier should be created by the overlapping of gown components (e.g., gloves overlapping sleeves). If an element of the gown is found to be torn or defective, it should be changed immediately.

PERSONNEL AND CLEANROOM CLOTHING: ISO 14644-5

ISO 14644-5 (2002) Cleanroom Operations notes in part 4.2 Cleanroom Clothing, that the environment and the product shall be protected from contamination generated by the personnel and their clothing. In order to maximize the containment, the choice of barrier fabric, the clothing style and the extent of coverage of personnel by the garment shall be established. It is also pointed out that cleanroom clothing shall be made of minimal linting fabrics and materials resisting breakdown and not shedding additional contamination. The necessary cleaning, processing and packaging shall be defined.

The informative Annex B gives guidance concerning the cleanroom requirements. It describes the prime function of the cleanroom clothing as a barrier filter that protects product and process from human contamination enveloping the person and not allowing significant amounts of unfiltered body emissions to be dispersed into the cleanroom. It is also noted that an effective cleanroom undergarment in combination with cleanroom clothing can give additional reduction in the dispersion from people.

The product cleanliness and process requirements affect the choice of cleanroom clothing but will in general consist of hoods, caps, helmets, coveralls, overboots, gloves, facemasks and goggles or safety glasses.

The Annex B gives information regarding the general choice of cleanroom clothing with regard to design and construction, fabrics and comfort aspects. Cleaning processes are not described here but final treatment, packaging, transports and storage of cleanroom clothing are discussed.

Annex C Personnel deals with training, access, clothing and personal items, hygiene, cleanroom clothing changing procedures, discipline and conduct, and safety. A typical procedure for changing into cleanroom clothing is outlined.

The information given in this ISO standard on cleanrooms and associated controlled environments is general and not adapted to the specific requirements of the pharmaceutical industry. However, it gives very useful information on the choice of appropriate clothing system and the necessary specifications for using, processing transporting and storing the cleanroom clothing systems.

SOME CALCULATIONS

TURBULENT MIXING AIR

It is possible to make a mathematical model of the level of airborne contaminants in a cleanroom with completely turbulent mixing air, if the contamination sources and the design of the ventilation system are known. Such descriptions of ventilation systems can be found in various text books.

With the assumption of no leakage into the room and the final filters having an efficiency close to 100% (HEPA-filters), the simplest possible expression describing the concentration in the room becomes:

$$c = \frac{q_s}{Q} \tag{1}$$

Where:
q_s = source strength, outward particle flow (number/s), bacteria-carrying particle (CFU/s)

Q = total air flow (m^3/s)

With Equation (1) some estimation is given in the two following examples:

Example 1

In an aseptic filling room of 75 cubic meter and 20 air changes per hour the only contamination sources are the cleanroom dressed operators. How many operators are allowed in the filling room if EU GMP requirements for Grade B shall be fulfilled?

Limit values of EU GMP Grade B in operation (see Guidelines for Pharmaceutical Production):

Number of particles ≥ 0.5 μm   350,000 per m^3

Number of particles ≥ 5 μm   2,000 per m^3

Number of airborne CFU   10 per m^3

The total air flow becomes:

$$Q = 75 \cdot \frac{20}{3600} \approx 0.42 \ \text{m}^3/\text{s}$$

Particles

Table 10 shows that maximum particle source strength for high quality cleanroom clothing system occurs around 25 washing/sterilizing cycles. For particles equal to and larger than 0.5 μm and particles equal to and larger than 5 μm the values are 3,950 particles per second and 70 particles per second, respectively. According to Equation (1) the concentrations become:

Particles ≥ 0.5 μm:

$$c = \frac{q_s}{Q} = \frac{3950}{0.42} \approx 9405 \ \text{particles/m}^3$$

Particles ≥ 5 μm:

$$c = \frac{q_s}{Q} = \frac{70}{0.42} \approx 167 \ \text{particles/m}^3$$

The maximum allowed number of people in the filling room depending on respective particle size will be the limit value of the EU GMP Grade B divided by the calculated concentration.

Particles ≥ 0.5 μm:

$$n_a = \frac{350000}{9405} \approx 37.2 \qquad \text{gives 37 operators}$$

Particles ≥ 5 µm

$$n_a = \frac{2000}{167} \approx 12.0 \qquad \text{gives 12 operators}$$

With regard to particles equal to and larger than 0.5 µm and particles equal to and larger than 5 µm the maximum allowed number of operators within the filling room are 37 and 12, respectively. It should be noted that the grade B demand for particles equal to and larger than 5 µm implies more strict limitations than that of particles equal to and larger than 0.5 µm.

CFU

Table 10 shows that the source strength for CFU increases with the number of washing/sterilizing cycles. Since the calculations follow the same steps that are described above for particles, initial data from Table 10 for CFU and high quality cleanroom clothing systems, and calculated results are summarized in Table 14.

Table 14. Source strengths (CFU per second) for high quality cleanroom clothing systems and estimated maximum number of operators allowed.

Number of washing/sterilizing cycles (number)	Source strength (CFU/s)	Concentration (Eq (1)) (CFU/m³)	Maximum number of operators (number)
1	0.38	0.90	11
25	0.49	1.17	8
50	1.14	2.71	3

Table 14 shows – with EU GMP grade B requirements – that after 50 washing/sterilizing cycles, the maximum number of operators allowed are three. Should the FDA Draft (2002, 2003) be used instead – with Clean Area Classification 1000 where the microbiological limit is set to 7 CFU per cubic meter – the maximum number of operators allowed would be only two.

Summary

The results of this example show, when the operators are the only contamination source, that the grade B limit value for CFU presents a stricter requirement than that of particles. On the other hand, it should be noted that in many production lines the machinery generate particles, which could be a reason for the larger ranges of Grade B limits for particles equal to and larger than 0.5 µm. If unidirectional flow units with HEPA-filtered air are used, the situation will be improved. The airflow through the units shall be added to the room air flow in Equation (1).

If, in the filling room in this example discussed here, a unidirectional flow unit with HEPA-filtered air is installed with the same air flow as the room air flow, the total air flow becomes 0.84 cubic meter per second. This means that the maximum number of allowed people in the filling room will be about twice those values, which are estimated in the above given example.

Example 2

How many people dressed in surgical clothing systems can be in an operating room with turbulent mixing air when the total air flow is 2,000 cubic meter per hour during

    a)   a conventional operation
    b)   an orthopedic operation (e.g., hip joint replacement).

Recommended values of CFU per m^3:

Conventional operation	<100 per m^3
Orthopedic operation	approximately 5 per m^3

The air flow is:

$$Q = \frac{2000}{3600} \approx 0.56 \ \text{m}^3\big/\text{s}$$

Table 10 shows source strength data for CFU and surgical clothing system washed once, 25 times and 50 times, respectively. The calculation procedure is the same as described in Example 1, and initial data and calculated results are directly given in Tables 15 and 16.

Table 15. Source strengths (CFU per second) for surgical clothing systems and estimated number of persons allowed during conventional operation.

Number of washing cycles (number)	Source strength (CFU/s)	Concentration (Eq(1)) (CFU/m^3)	Maximum number of operators (number)
1	1.7	3.04	32
25	4.2	7.5	13
50	9.0	16.1	6

Table 16. Source strengths (CFU per second) for surgical clothing systems and estimated number of persons allowed during orthopedic operation (e.g., hip joint replacement).

Number of washing cycles (number)	Source strength (CFU/s)	Concentration (Eq(1)) (CFU/m^3)	Maximum number of operators (number)
1	1.7	3.04	1
25	4.2	7.5	<1
50	9.0	16.1	<1

The results indicate that surgical clothing systems for conventional operations should not be used more than 50 washing cycles. For orthopedic operations the surgical clothing system discussed here should not be used at all. It should be mentioned that hip joint replacement operations are often performed in a HEPA-filtered unidirectional air flow system with the surgical team dressed in high quality clothing systems. The mathematical expression for an operator in a unidirectional air flow will be discussed in the following part.

CLEANROOM DRESSED OPERATOR IN UNIDIRECTIONAL AIR FLOW

The purpose of this part is to present a mathematical model describing the dispersion of airborne contaminants when a cleanroom dressed operator is standing in vertical unidirectional air flow.

A vertically standing cylinder with its axis in a vertical position and with a constant radius approximates the operator. A solution for a continuous finite cylinder source has been reported by Ljungqvist et al. (2003). The center-axis of the cylinder is situated in the same position as the positive part of the x-axis with the origin situated on the center-axis at one side of the cylinder. The concentration in a parallel flow with constant velocity, v_0, in the x-direction, when $(\partial c/\partial t = 0)$ can, in cylindrical coordinates $((\rho, x)$ with symmetry, i.e., φ-independence), be expressed as:

$$
c(\rho,x) = \frac{q_l}{2 \cdot \pi \cdot \sqrt{v_0 \cdot D \cdot \rho \cdot \rho_1}} \cdot \left\{ 2 \cdot \sqrt{x} \cdot \text{ierfc}\left[\sqrt{\frac{v_0}{D}} \cdot (\rho - \rho_1) \cdot \frac{1}{2 \cdot \sqrt{x}} \right] - \right.
$$

$$
\frac{1}{2 \cdot \sqrt{\dfrac{v_0}{D}}} \cdot \left(\frac{1}{\rho} + \frac{3}{\rho_1} \right) \cdot x \cdot \text{i}^2\text{erfc}\left[\sqrt{\frac{v_0}{D}} \cdot (\rho - \rho_1) \cdot \frac{1}{2 \cdot \sqrt{x}} \right] + \qquad (2)
$$

$$
\left. \frac{3}{16 \cdot \dfrac{v_0}{D}} \cdot \left(\frac{3}{\rho^2} + \frac{2}{\rho \cdot \rho_1} + \frac{11}{\rho_1^2} \right) \cdot x \cdot \sqrt{x} \cdot \text{i}^3\text{erfc}\left[\sqrt{\frac{v_0}{D}} \cdot (\rho - \rho_1) \cdot \frac{1}{2 \cdot \sqrt{x}} \right] \right\}
$$

where:

q_l	=	outward particle flow per unit of length from the cylinder source (number/(s, m))
v_0	=	constant velocity in the x-direction (m/s)
D	=	diffusion coefficient (m²/s)
ρ_1	=	radius of the cylinder (m)
i^nerfc	=	repeated integrals of the complementary error function (erfc)

Numerical calculations show that the concentration in Equation (2), if only the two first terms are included, the accuracy will be about 3%; and if only the first term is used the accuracy decreases to 15%.

Equation (2) can be used to estimate the concentration profile when a cleanroom dressed operator is standing in a vertical unidirectional flow of HEPA-filtered air. The operator is approximated by a vertical standing cylinder with its center-axis in a vertical position along the x-axis and with a constant radius ρ_1. The origin is situated in the center of the top of the cylinder (operator's head). The unidirectional air flow with constant velocity, v_0, is moving downwards in the x-direction and particles are emitted from the cylinder surface.

Data from Figure 13 show that the strength of the contamination source, people during activity dressed in modern cleanroom clothing systems, for particles equal to and larger than 0.5 μm is estimated to be 500 to 800 particles per second and meter.

Numerical data given by Ljungqvist (1979), ($v_0 = 0.45$ m/s and $D = 2.4 \cdot 10^{-4}$ m^2/s) and the value of the cylinder radius chosen to 0.15 meter seems to be representative. With the above given numerical values of the concentration profile along the air stream at a distance of 0.1 meter from the cylinder surface, i.e., a distance of 0.25 meter from center of the cylinder ($\rho = 0.25$ m) is shown in Figure 28.

Figure 28. Concentration (number of particles $\geq$ 0.5 μm per m^3) at a distance of 0.1m (ρ = 0.25 m) from the cylinder surface (with ρ_1 = 0.15 m) as a function of the distance (x-direction) from the top of the cylinder in a unidirectional air flow with the velocity 0.45 m/s and the diffusion coefficient 2.4 $\times$ 10^{-4} m^2/s. The source strength of particles $\geq$ 0.5 μm per second and meter is chosen to 800.

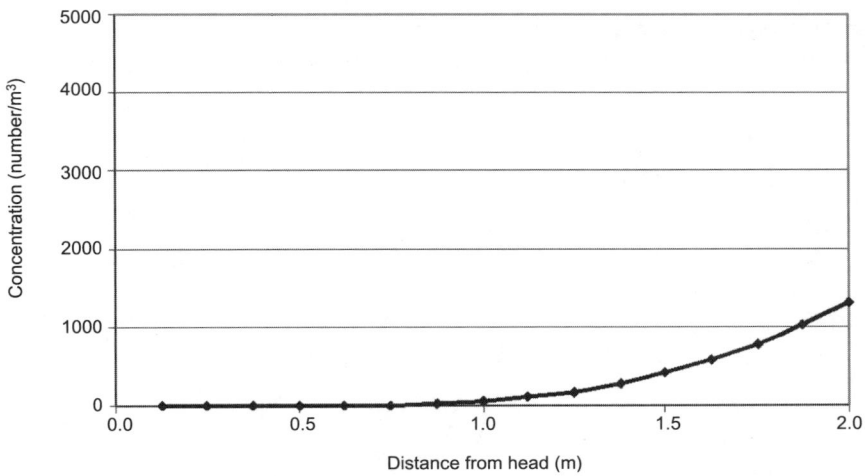

Figure 28 shows that the concentration increases with increasing distance of x. At distances less than 0.7 meter (head and chest region) the values are small but at distances larger than 1.5 meter (below knees) the values increases rapidly. Further more, results from calculations of Equation (2) give that the concentration increases with decreasing velocity.

In factual situations, within the region close to the floor ($x \geq 1.5$ meter) the air flow is no longer unidirectional, as the air flow in most cases deviates to exhaust systems in the walls. This means that aseptic work within the region close to the floor always constitutes a risk situation and should be avoided.

Experimental tests have been performed by Jonsson (2002) in the dispersal chamber at KTH with a cylindrical source. The particle counter used was sensitive to particle concentrations exceeding 10^6 particles (≥ 0.5 µm) per cubic feet. Therefore the measurements have been carried out to find the distance from the center of the cylinder (ρ) to a position where the particle concentration is close to zero. This is called contamination region edge.

The experimentally-received results verify the theoretical expression in Equation (2), while there is a correlation between measured and theoretical calculated values.

Figure 29. Contamination region edge (c (ρ, x) $\approx$ 35 particles (≥ 0.5 µm)per m³) at the air velocity 0.45 m/s. Thicker line represent edge for measured values. Thinner lines show the position of the calculated edges according to the theoretical model and the source strengths 500 and 20,000 particles ≥ 0.5 µm per meter and second, respectively.

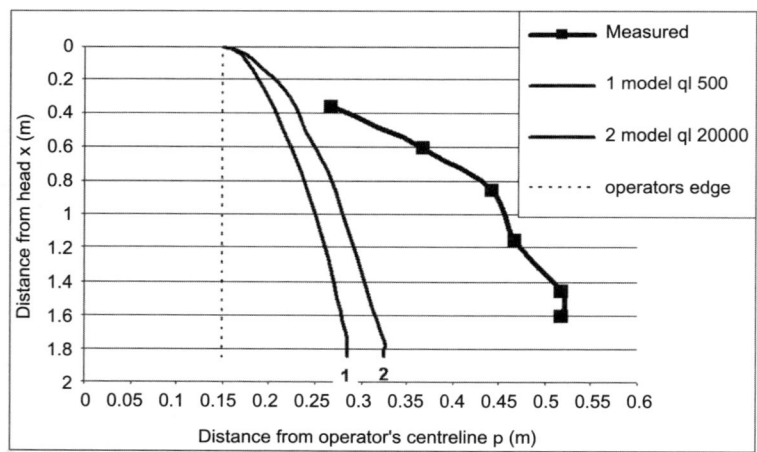

Measurements have also been performed on an operator inside the dispersal chamber at an air velocity of 0.45 meter per second. The operator has the height of 1.86 meter and an estimated radius of 0.15 meter. The measurements have been performed with the operator's arms moving in a calm manner in standard cycles (one hand from hip to shoulder and back, then the same movement with the other hand), see Jonsson (2002). The source strength has been measured and calculated to 500 particles (≥ 0.5 μm) per second and meter. This value might be somewhat small so an additional value of the source strength chosen to 20,000 particles (≥ 0.5 μm) per second and meter has been used in the theoretical calculation. The data in Table 9 show that this value of the source strength is comparable with "lab coat – disp coat". Figure 29 shows both the theoretical and measured contamination region edges for an operator. The contamination region edge is here set to 35 particles (≥ 0.5 μm) per cubic meter (1 particle per ft^3).

Figure 29 shows that the measured results with an operator in activity (arm movements) deviate from results based on the mathematical model. The distances from the center of the cylinder to the contamination region edge are about 50% larger for values measured with moving operator than those theoretically calculated.

It should be noted that in spite of the large difference in source strengths between the two theoretically calculated curves (predicting operator standing still) in Figure 29, the contamination regions differ less than 0.1 meter.

REFERENCES

Aczel, A. D. (1999) *Complete Business Statistics*, Mc Graw-Hill Book Co, Singapore.

Bradley, A., Probert, S. P., Sinclair, S. C., and Tallentyre, A. (1991) Airborne Microbial Challenges of Blow/Fill/Seal Equipment; A case Study, *Journal of Parenteral Science and Technology*, Vol. 45, No. 4, pp. 187–192.

Daniel, W. W. (1974) *Biostatistics: A Foundation for Analysis in Health Sciences*, John Wiley & Sons Inc., New York.

EU GMP (1997) European Commission, The Rules Governing Medicinal products in the European Community, Vol. IV, Guide to Good Manufacturing Practice, Annex 1: Manufacture of sterile medicinal products (revised 2003).

FDA (1987) Food and Drug Administration, Guideline on Sterile Drug Products Produced by Aseptic Processing, Center for Drugs and Biologics and Office of Regulatory Affairs, Rockville, Maryland.

FDA (1996) Food and Drug Administration, CGMP Regulations, 210/211, U.S. Government Printing Office.

FDA (2002) Food and Drug Administration, Sterile Drug Products Produced by Aseptic Processing, Draft, Rockville, Maryland.

FDA (2003) Food and Drug Administration, Guidance for Industry, Sterile Drug Products Produced by Aseptic Processing – Current Good Manufacturing Practice, Draft Guidance, Rockville, Maryland.

Gustavsson, J. (1999) *Clean room filters – a guide*, Technical information, 990525, Camfil AB, Trosa.

Hoborn, J. (1981) *Humans as Dispersers of Microorganisms – Dispersion Pattern and Prevention*, Ph.D. Thesis, Dep. of Clinical Bacteriology, Inst. of Medical Microbiology, University of Göteborg, Sweden.

ISO 14644-1 (1999) Cleanrooms and associated controlled environments, Part 1: Classification of air cleanliness.

ISO 14644-4 (2001) Cleanrooms and associated controlled environments, Part 4: Design, construction and start-up.

ISO 14644-5 (2002) Cleanrooms and associated controlled environments, Part 5: Cleanroom Operations.

ISO/FDIS 14698-1 (2003) Cleanrooms and associated controlled environments – Biocontamination Control, Part 1: General principles and methods.

IES-RP-CC001.3 (1992) *HEPA and ULPA Filters*, IES Institute of Environmental Sciences, Illinois.

IES-RP-CC003.2 (1993) *Garment System Considerations for Cleanroom and Other Controlled Environments*, IES Institute of Environmental Sciences, Illinois.

Jonsson, R. (2002) *Particle dispersal from a cylindrical source in a clean zone with unidirectional air flow*, M.Sc. Thesis, report No. 85, Building Services Engineering, KTH, Stockholm (in Swedish).

Ljungqvist, B. (1979) *Some observations on the interaction between air movements and the dispersion of pollution*; Ph.D. Thesis, KTH, Swedish Council for Building Research, Document D8:1979, Stockholm.

Ljungqvist, B., and Reinmüller, B. (1995) Hazard Analyses of Airborne Contamination in Clean Rooms - Application of a Method for Limitation of Risks, *PDA Journal of Pharmaceutical Science and Technology*; 49, pp. 239–243.

Ljungqvist, B., and Reinmüller, B. (1998) Active Sampling of Airborne Viable Particles in Controlled Environments: a comparative study of common instruments, *European Journal of Parenteral Sciences*, 3 (3), pp. 59–62.

Ljungqvist, B., and Reinmüller, B. (2003a) *People as a Contamination Source; Clean room clothing after 25 and 50 washing/sterilization cycles*, Bulletin No. 60, Building Services Engineering, KTH, Stockholm.

Ljungqvist, B., and Reinmüller, B. (2003b) People as a contamination source; cleanroom clothing after 1, 25 and 50 washing/sterilizing cycles, *European Journal of Parenteral & Pharmaceutical Sciences*, 8 (3), pp. 75–80.

Ljungqvist, B., Reinmüller, B., Söderström, O. (2003) Cleanroom-dressed operator in unidirectional airflow; a mathematical model of contamination risks, *European Journal of Parenteral & Pharmaceutical Sciences*, 8 (1), pp. 11–14.

Moore, D. S. (1997) *Statistics, Concepts and Controversies*, W.H. Freeman and Company, New York.

PQRI (2003) PQRI *Recommendation on Aseptic processing:* Final Report, http//www.pqri.org/aseptic/asepticsurvey.htm (March 2003).

Reinmüller, B. (2001) *Dispersion and Risk Assessment of Airborne Contaminants in Pharmaceutical Clean Rooms*, Ph.D. Thesis, Bulletin No. 56, Building Services Engineering, KTH, Stockholm.

Reinmüller, B., and Ljungqvist, B. (2000) Evaluation of cleanroom garments in a dispersal chamber - some observations, *European Journal of Parenteral Sciences*, 5(3), pp. 55–58.

Reinmüller, B., and Ljungqvist, B. (2002) *People as a Contamination Source; Modern clean room clothing systems*, Bulletin No. 57, Building Services Engineering, KTH, Stockholm.

Reinmüller, B., and Ljungqvist, B. (2003) Modern Cleanroom Clothing Systems; People as a Contamination Source, *PDA Journal of Pharmaceutical Science and Technology*, Vol. 57 No. 3, pp. 114–125.

Tammelin, A., Domicel, P., Hambraeus, A., Stråhle, E. (2000) Dispersal of methicillin-resistant *Staphylococcus epidermidis* by staff in an operating suite for thoracic and cardiovascular surgery: relation to skin carriage and clothing, *Journal of Hospital Infection*, 44, pp. 119–126.

USP 27-NF 22 (2004) <1116> Microbiological evaluation of clean rooms and other controlled environments, United States Pharmacopeial Convention, Rockville, MD.

Whyte, W. (1999) Cleanroom Design, 2nd Ed., edited by Whyte, John Wiley & Sons, Chichester.

Whyte, W., and Bailey, P. (1985) Reduction of Microbial Dispersion by Clothing, *Journal of Parenteral Science and Technology*; 39, pp. 51–60.

Whyte, W., Vesley, D., Hodgson, R. (1976) Bacterial dispersion to operating room clothing, *J. Hyg., Camb.*, 76, pp. 367–378.

Ytterman, P. (1998) *Qualification of modified body-box for evaluation of cleanroom garments* M.Sc. Thesis, Report No. 59, Building Services Engineering, KTH, (in Swedish).

SYMBOLS AND ABBREVIATIONS

ANOVA   Analysis of Variance

c Concentration; particles, number/m^3, bacteria-carrying particles, CFU/m^3

CFU   Colony Forming Unit; bacteria-carrying particles, number

cGMP   Current Good Manufacturing Practice

D Diffusion coefficient; m^2/s

ECF   Electro-Conductive Fibers

FDA   Food & Drug Administration (US)

GMP   Good Manufacturing Practice

GMV   Grand Mean Values

HEPA   High Efficiency Particulate Air

MV   Mean Values

n_a Number of maximum allowed operators/persons in a room; number

q_l Outward particle flow from a cylinder per unit length; number/(s, m)

q_s Source strength, outward flow; particles, number/s; bacteria-carrying particles; CFU/s

Q Total air flow; m^3/s

RH   Relative Humidity

TSA   Tryptic Soy Agar

UDF   Unidirectional Flow

ULPA   Ultra Low Particulate Air

v_0 Constant air velocity in the x-direction; m/s

x Positional coordinate

ρ, φ Radius and angle in cylindrical coordinates

ρ_1 Radius of cylinder; m

APPENDIX: CLOTHING SYSTEMS EVALUATED

TEST SERIES 1

CLOTHING SYSTEMS USED

Reference clothing (Ref)

Head cover	No	
Face cover	No	
Coat/coverall	No	
Trousers	Jeans	Cotton
Blouse	T-shirt	Cotton
Gloves	No	
Shoes	Sandal	Plastic
Overshoes	No	
Underwear	Pants	Cotton
Socks	Tube socks	Cotton

Disposable coat (Disp)

Head cover	No	
Face cover	No	
Coat/coverall	Disposable coat	Non-woven surgical gown, cuffs at neck and sleeves and tied in the back (Baxter Surgical Gown)
Trousers	Jeans	Cotton
Blouse	T-shirt	Cotton
Gloves	No	
Shoes	Sandal	Plastic
Overshoes	No	
Underwear	Pants	Cotton
Socks	Tube socks	Cotton

Cleanroom clothing system (Cleanroom clothing)

Head cover	Cleanroom hood	100% polyester, Selguard®, washed and autoclaved several times
Face cover	Disposable sterile mask	BCR mask, Berkshire
Coat/coverall	Clean room coverall	100% polyester, Selguard®, washed and autoclaved several times
Trousers	No	
Blouse	No	
Gloves	Latex gloves	Sterile
Shoes	Sandal	Plastic
Overshoes	Cleanroom long boots	100% polyester, Selguard®, washed and autoclaved several times
Underwear	Cleanroom T-shirt and pants	Cotton
Socks	Tube socks	Cotton

TEST SERIES 2

PART 1: CLOTHING SYSTEMS USED

Cleanroom clothing system (without goggles)

Head cover	Disposable head cover	
	Tight cleanroom hood, sterile	100% polyester, Selguard®, washed and autoclaved once
Face cover	Disposable sterile mask	BCR mask, Berkshire
Coat/coverall	Cleanroom coverall, sterile	100% polyester, Selguard®, washed and autoclaved once
Trousers	No	
Blouse	No	
Gloves	Latex gloves	Sterile
Shoes	Sandal	Plastic
Overshoes	Cleanroom long boots, sterile	100% polyester, Selguard®, washed and autoclaved once
Underwear	Cleanroom shirt, pants	100% polyester Integrity® 1600 and cotton
Socks	Cleanroom socks	100% polyester

Cleanroom clothing system (goggles)

Head cover	Disposable head cover	
	Tight cleanroom hood, sterile	100% polyester, Selguard®, washed and autoclaved once
Face cover	Disposable sterile mask	BCR mask, Berkshire
	Goggles, disinfected	
Coat/coverall	Cleanroom coverall, sterile	100% polyester, Selguard®, washed and autoclaved once
Trousers	No	
Blouse	No	
Gloves	Latex gloves	Sterile
Shoes	Sandal	Plastic
Overshoes	Cleanroom long boots, sterile	100% polyester, Selguard®, washed and autoclaved once
Underwear	Cleanroom shirt, pants	100% polyester Integrity® 1600 and cotton
Socks	Cleanroom socks	100% polyester

Cleanroom clothing system (large mask)

Head cover	Disposable head cover	
	Tight cleanroom hood, sterile	100% polyester, Selguard®, washed and autoclaved once
Face cover	Disposable sterile mask	Large BCR mask, Berkshire
Coat/coverall	Cleanroom coverall, sterile	100% polyester, Selguard®, washed and autoclaved once
Trousers	No	
Blouse	No	
Gloves	Latex gloves	Sterile
Shoes	Sandal	Plastic
Overshoes	Cleanroom long boots, sterile	100% polyester, Selguard®, washed and autoclaved once
Underwear	Cleanroom shirt, pants	100% polyester Integrity® 1600 and cotton
Socks	Cleanroom socks	100% polyester

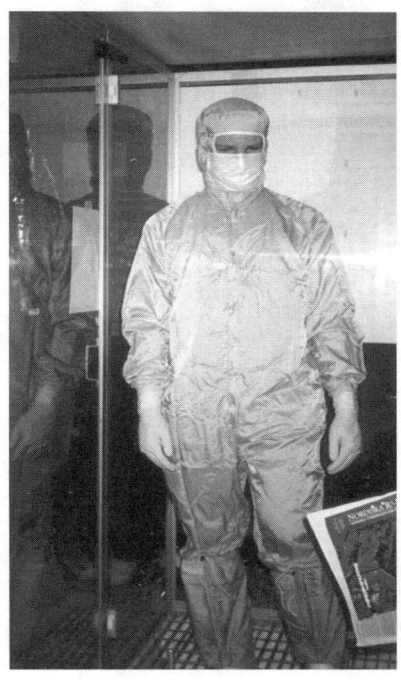

Cleanroom clothing system (mask)

Head cover	Disposable head cover	
	Tight cleanroom hood, sterile	100% polyester, Selguard®, washed and autoclaved once
Face cover	Disposable sterile mask	PCU 2000
Coat/coverall	Cleanroom coverall, sterile	100% polyester, Selguard®, washed and autoclaved once
Trousers	No	
Blouse	No	
Gloves	Latex gloves	Sterile
Shoes	Sandal	Plastic
Overshoes	Cleanroom long boots, sterile	100% polyester, Selguard®, washed and autoclaved once
Underwear	Cleanroom shirt, pants	100% polyester Integrity® 1600 and cotton
Socks	Cleanroom socks	100% polyester

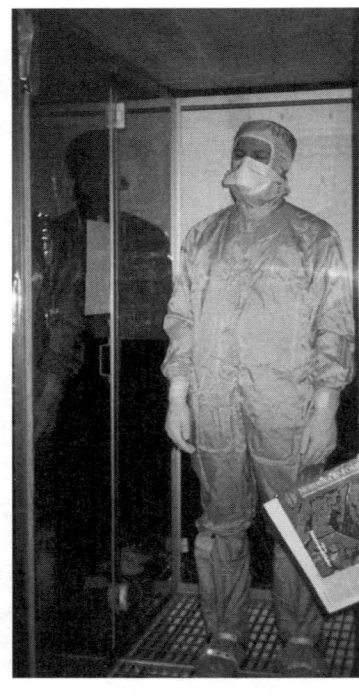

Cleanroom clothing system (tight hood)

Head cover	Disposable head cover	
	Tight cleanroom hood, sterile	100% polyester, Selguard®, washed and autoclaved once
Face cover	Disposable sterile mask	CR 45250 Berkshire
Coat/coverall	Cleanroom coverall, sterile	100% polyester, Selguard®, washed and autoclaved once
Trousers	No	
Blouse	No	
Gloves	Latex gloves	Sterile
Shoes	Sandal	Plastic
Overshoes	Cleanroom long boots, sterile	100% polyester, Selguard®, washed and autoclaved once
Underwear	Cleanroom shirt, pants	100% polyester Integrity® 1600 and cotton
Socks	Cleanroom socks	100% polyester

Cleanroom clothing system (loose hood)

Head cover	Disposable head cover	
	Loose cleanroom hood, sterile	100% polyester, Selguard®, washed and autoclaved once
Face cover	Disposable sterile mask	CR 45250 Berkshire
Coat/coverall	Cleanroom coverall, sterile	100% polyester, Selguard®, washed and autoclaved once
Trousers	No	
Blouse	No	
Gloves	Latex gloves	Sterile
Shoes	Sandal	Plastic
Overshoes	Cleanroom long boots, sterile	100% polyester, Selguard®, washed and autoclaved once
Underwear	Cleanroom shirt, pants	100% polyester Integrity® 1600 and cotton
Socks	Cleanroom socks	100% polyester

PART 2: CLOTHING SYSTEMS USED

Cleanroom clothing system (pharm)

Head cover	Disposable head cover	
Face cover	Disposable sterile mask	BCR mask, Berkshire
Coat/coverall	No	
Trousers	Cleanroom trousers, cuffs at legs	35% cotton and 65% polyester, washed once
Blouse	Cleanroom blouse, cuffs at sleeves and neck, buttons up front	35% cotton and 65% polyester, washed once
Gloves	Latex gloves	Sterile
Shoes	Sandal	Plastic
Overshoes	No	
Underwear	Cleanroom T-shirt and pants	Cotton
Socks	Cleanroom socks	100% polyester

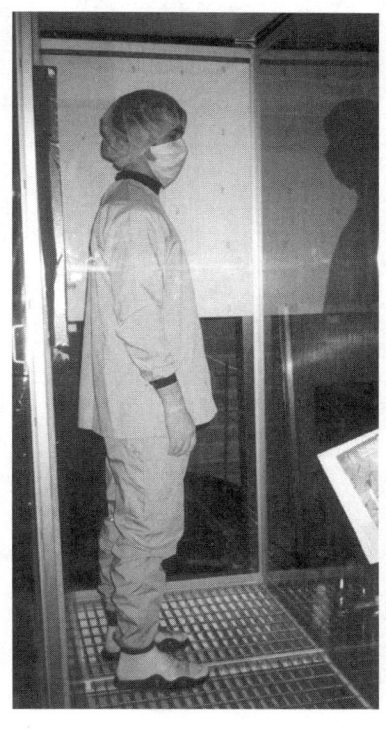

Surgical cleanroom clothing system (surgeon)

Head cover	Disposable head cover	
Face cover	Disposable sterile mask	BCR mask, Berkshire
Coat/coverall	No	
Trousers	Cleanroom trousers with cuffs at legs	50% polyester and 50% cotton, Cambric, washed once
Blouse	Cleanroom blouse, short sleeves, tucked inside trousers	50% polyester and 50% cotton, Cambric, washed once
Gloves	Latex gloves	Sterile
Shoes	Sandal	Plastic
Overshoes	No	
Underwear	Cleanroom T-shirt and pants	Cotton
Socks	Cleanroom socks	100% polyester

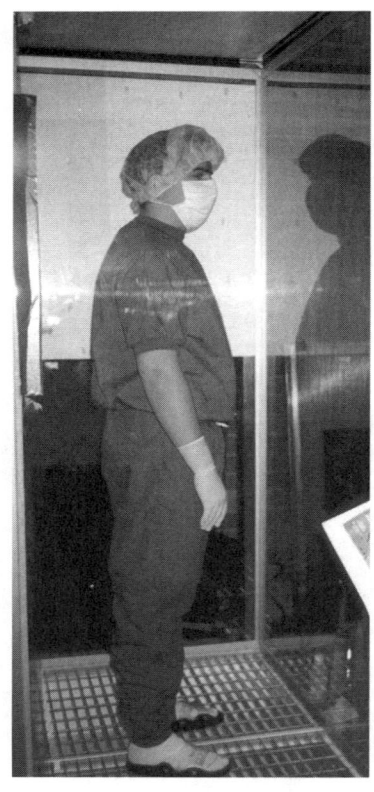

Cleanroom clothing system (Coverall Japan)

Head cover	Disposable head cover	
	Tight cleanroom hood, sterile	100% polyester, Selguard®, washed and autoclaved once
Face cover	Disposable sterile mask	BCR mask, Berkshire
Coat/coverall	Cleanroom coverall, sterile	100% polyester, Selguard®, washed and autoclaved once
Trousers	No	
Blouse	No	
Gloves	Latex gloves	Sterile
Shoes	Sandal	Plastic
Overshoes	Cleanroom long boots, sterile	100% polyester, Selguard®, washed and autoclaved once
Underwear	Cleanroom underwear, long-legged trousers and long-sleeved shirt	100% polyester Integrity® 1600
Socks	Cleanroom socks	100% polyester

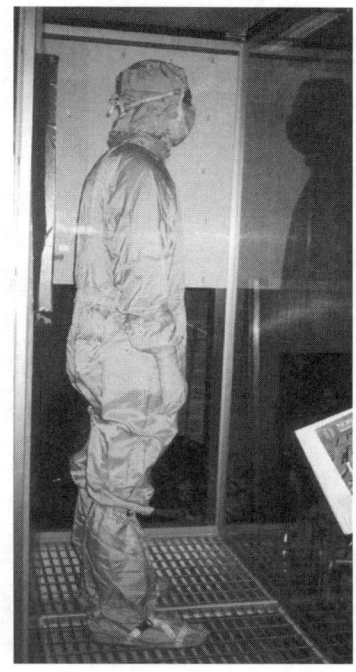

Cleanroom clothing (Coverall US)

Head cover	Disposable head cover	
	Tight cleanroom hood, sterile	100% polyester, Integrity® 1800, washed and autoclaved once
Face cover	Disposable sterile mask	BCR mask, Berkshire
Coat/coverall	Cleanroom coverall, sterile	100% polyester, Integrity® 1800, washed and autoclaved once
Trousers	No	
Blouse	No	
Gloves	Latex gloves	Sterile
Shoes	Sandal	Plastic
Overshoes	Cleanroom long boots, sterile	100% polyester, Integrity® 1800, washed and autoclaved once
Underwear	Cleanroom underwear, long-legged trousers and long-sleeved shirt	100% polyester Integrity® 1600
Socks	Cleanroom socks	100% polyester

Cleanroom clothing (Tyvek®)

Head cover	Disposable head cover	
Face cover	Disposable sterile mask	BCR mask, Berkshire
Coat/coverall	Disposable cleanroom coverall with hood, sterile	Non-woven material, 100% polyethylene, Tyvek®
Trousers	No	
Blouse	No	
Gloves	Latex gloves	Sterile
Shoes	Sandal	Plastic
Overshoes	Cleanroom long boots, sterile	Non-woven material, 100% polyethylene, Tyvek®
Underwear	Cleanroom T-shirt and pants	Cotton
Socks	Cleanroom socks	100% polyester

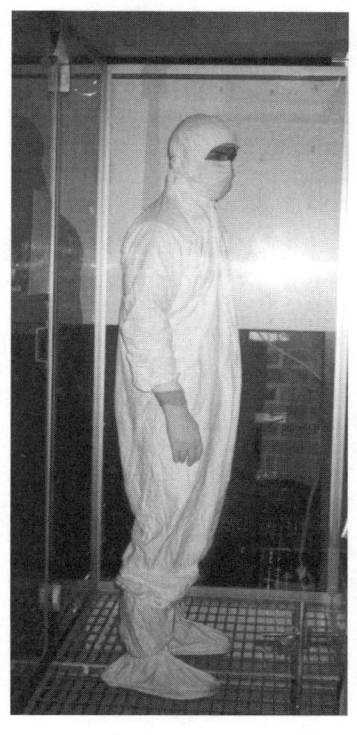

TEST SERIES 3

CLOTHING SYSTEMS 25 WASHES

Surgical cleanroom clothing system (surgeon)

Head cover	Disposable head cover	
Face cover	Disposable sterile mask	BCR mask, Berkshire
Coat/coverall	No	
Trousers	Cleanroom trousers with cuffs at legs	50% polyester and 50% cotton, Cambric
Blouse	Cleanroom blouse, short sleeves, tucked inside trousers	50% polyester and 50% cotton, Cambric
Gloves	Latex gloves	Sterile
Shoes	Sandal	Plastic
Overshoes	No	
Underwear	Cleanroom T-shirt and pants	Cotton
Socks	Cleanroom socks	100% polyester

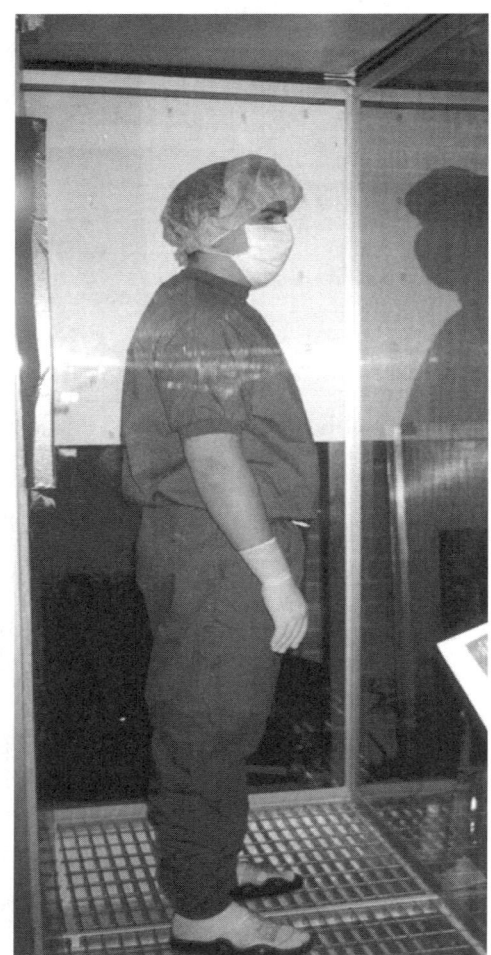

Cleanroom clothing system (Coverall Japan; J+underwear+autocl.)

Head cover	Disposable head cover and	
	Tight cleanroom hood, sterile	Polyester, Selguard®, twills; 25 washes/autocl
Face cover	Disposable sterile mask	BCR mask, Berkshire
Coat/coverall	Cleanroom coverall, sterile	Polyester, Selguard®, twills; 25 washes/autocl
Trousers	No	
Blouse	No	
Gloves	Latex gloves	Sterile
Shoes	Sandal	Plastic
Overshoes	Cleanroom long boots, sterile	Polyester, Selguard®, twills; 25 washes/autocl
Underwear	Cleanroom underwear, long-legged trousers and long-sleeved shirt	Polyester, Integrity® 1600, plain weave
Socks	Cleanroom socks	100% polyester

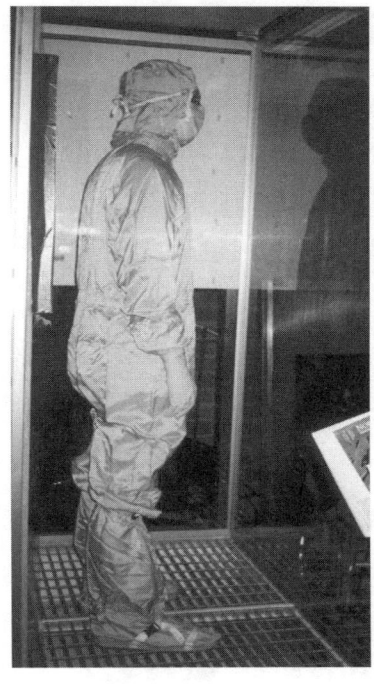

Cleanroom clothing system (Coverall Japan; J+shirt+autocl.)

Head cover	Disposable head cover and Tight cleanroom hood	Polyester, Selguard®, twills; 25 washes/autocl
Face cover	Disposable sterile mask	BCR mask, Berkshire
Coat/coverall	Cleanroom coverall, sterile	Polyester, Selguard®, twills; 25 washes/autocl
Trousers	No	
Blouse	No	
Gloves	Latex gloves	Sterile
Shoes	Sandal	Plastic
Overshoes	Cleanroom long boots, sterile	Polyester, Selguard®, twills; 25 washes/autocl
Underwear	Cleanroom underwear; long-sleeved shirt	Polyester, Integrity® 1600, plain weave
	Pants	Cotton
Socks	Cleanroom socks	100% polyester

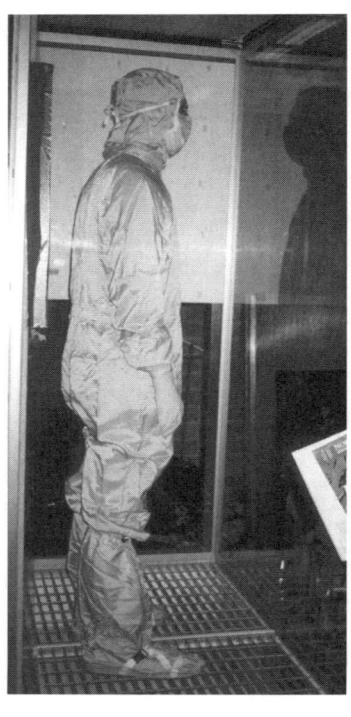

Cleanroom clothing (Coverall US; US+underwear+autocl.)

Head cover	Disposable head cover	
	Tight cleanroom hood, sterile	Polyester, Integrity® 1800, plain weave; 25 washes/autocl
Face cover	Disposable sterile mask	BCR mask, Berkshire
Coat/coverall	Cleanroom coverall, sterile	Polyester, Integrity® 1800, plain weave; 25 washes/autocl
Trousers	No	
Blouse	No	
Gloves	Latex gloves	Sterile
Shoes	Sandal	Plastic
Overshoes	Cleanroom long boots, sterile	Polyester, Integrity® 1800, plain weave; 25 washes/autocl
Underwear	Cleanroom underwear, long-legged trousers and long-sleeved shirt	Polyester Integrity® 1600, plain weave
Socks	Cleanroom socks	100% polyester

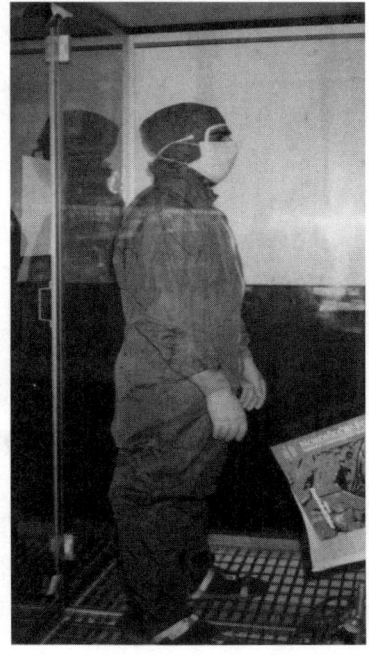

CLOTHING SYSTEMS 50 WASHES

Surgical cleanroom clothing system (surgeon)

Head cover	Disposable head cover	
Face cover	Disposable sterile mask	BCR mask, Berkshire
Coat/coverall	No	
Trousers	Cleanroom trousers with cuffs at legs	50% polyester and 50% cotton, Cambric
Blouse	Cleanroom blouse, short sleeves, tucked inside trousers	50% polyester and 50% cotton, Cambric
Gloves	Latex gloves	Sterile
Shoes	Sandal	Plastic
Overshoes	No	
Underwear	Cleanroom T-shirt and pants	Cotton
Socks	Cleanroom socks	100% polyester

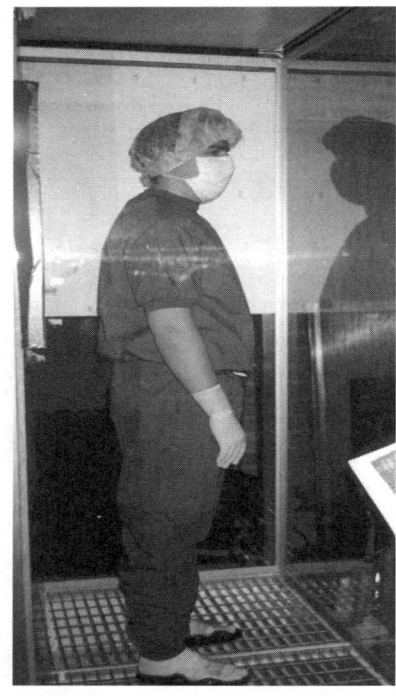

Cleanroom clothing system (Coverall Japan; J+underwear)

Head cover	Disposable head cover and Tight cleanroom hood, sterile	Polyester, Selguard®, twills
Face cover	Disposable sterile mask	BCR mask, Berkshire
Coat/coverall	Cleanroom coverall	Polyester, Selguard®, twills
Trousers	No	
Blouse	No	
Gloves	Latex gloves	Sterile
Shoes	Sandal	Plastic
Overshoes	Cleanroom long boots	Polyester, Selguard®, twills
Underwear	Cleanroom underwear, long-legged trousers and long-sleeved shirt	Polyester, Integrity® 1600, plain weave
Socks	Cleanroom socks	100% polyester

Cleanroom clothing system (Coverall Japan; J+shirt+autocl.)

Head cover	Disposable head cover and Tight cleanroom hood	Polyester, Selguard®, twills; 50 washes/autocl
Face cover	Disposable sterile mask	BCR mask, Berkshire
Coat/coverall	Cleanroom coverall, sterile	Polyester, Selguard®, twills; 50 washes/autocl
Trousers	No	
Blouse	No	
Gloves	Latex gloves	Sterile
Shoes	Sandal	Plastic
Overshoes	Cleanroom long boots, sterile	Polyester, Selguard®, twills; 50 washes/autocl
Underwear	Cleanroom underwear; long-sleeved shirt	Polyester, Integrity® 1600, plain weave
	Pants	Cotton
Socks	Cleanroom socks	100% polyester

Cleanroom clothing system (Coverall Japan; J+underwear+autocl.)

Head cover	Disposable head cover and Tight cleanroom hood, sterile	Polyester, Selguard®, twills; 50 washes/autocl
Face cover	Disposable sterile mask	BCR mask, Berkshire
Coat/coverall	Cleanroom coverall, sterile	Polyester, Selguard®, twills; 50 washes/autocl
Trousers	No	
Blouse	No	
Gloves	Latex gloves	Sterile
Shoes	Sandal	Plastic
Overshoes	Cleanroom long boots, sterile	Polyester, Selguard®, twills; 50 washes/autocl
Underwear	Cleanroom underwear, long-legged trousers and long-sleeved shirt	Polyester, Integrity® 1600, plain weave
Socks	Cleanroom socks	100% polyester

Cleanroom clothing (Coverall US; US+underwear+autocl.)

Head cover	Disposable head cover Tight cleanroom hood, sterile	Polyester, Integrity® 1800, plain weave; 50 washes/autocl
Face cover	Disposable sterile mask	BCR mask, Berkshire
Coat/coverall	Cleanroom coverall, sterile	Polyester, Integrity® 1800, plain weave; 50 washes/autocl
Trousers	No	
Blouse	No	
Gloves	Latex gloves	Sterile
Shoes	Sandal	Plastic
Overshoes	Cleanroom long boots, sterile	Polyester, Integrity® 1800, plain weave; 50 washes/autocl
Underwear	Cleanroom underwear, long-legged trousers and long-sleeved shirt	Polyester Integrity® 1600, plain weave
Socks	Cleanroom socks	100% polyester

SUPPLIER'S SPECIFICATIONS

COVERALLS

Coverall Japan

Material	Selguard Continuous filament polyester fabric with an ESD stripe
Type of weave	3/1 twill
Weight	120 gram per m^2
Density: warp x fill	190 x 190 threads per inch
Dernier: warp x fill	75 x 100
Filtration Efficiency* (Particles $\geq$ 0.3 µm)	85% IES RP-CC-003-87-T (modified)
Filtration Efficiency* (Particles $\geq$ 0.5 µm)	88% IES RP-CC-003-87-T (modified)
Moist. Vapor T (MVT)* (g/m^2/1 hour)	290 at 40 ºC, 90% RH JIS L-1099,4-1-1, A1 (ref. ASTM 96-1990/B)

* Fabric tested after zero wash cycles

Coverall US

Material	Integrity® 1800 Continuous filament polyester fabric with an ESD stripe
Type of weave	Plain
Weight	95 gram per m^2
Mean Pore Size	3 micron
Filtration Efficiency (Particles $\geq$ 0.5 µm)	92% IEST-RP-CC003.2
Moist. Vapor T (MVT) ($g/m^2/24$ hour)	1100 ASTM D-96 Method B at 50% RH
Antimicrobial	Yes Method: Bromo Phenyl Blue

Data is represented as test results and is not to serve as specifications.

Coverall Tyvek®

Material	Non-woven Polyethylene

CLEANROOM UNDERWEAR

Material	Integrity® 1600 Continuous filament polyester fabric with an ESD stripe
Type of weave	Plain
Weight	90 gram per m^2
Mean Pore Size	30 micron
Filtration Efficiency (Particles $\geq$ 0.5 µm)	Not Applicable
Moist. Vapor T (MVT) ($g/m^2/24$ hour)	1200 ASTM D-96 Method B at 50% RH
Antimicrobial	Yes Method: Bromo Phenyl Blue

Data is represented as test results and is not to serve as specifications.

Masks

Material; Inner layer	Non-woven polypropylene, white
Material; Outer layer	Non-woven polypropylene, light blue
Filter Media	Melt-down Non-woven Polypropylene
Ear-loops	Polyurethane
Side Tapes	Non-woven Polyester
Nose Clamp	Vinyl Covered steel wire
Particle Filtration Efficiency % (0.1 μm)	96.3
Bacterial Filtration efficiency % (3.0 μm)	99.0
Particle/ft^3 (HelmkeDrum)	150
Sterilization Method	Gamma irradiation ≥ 25 kGy

Gloves

Material	Natural Rubber Latex
Features	Powder free, hand specific, anatomical shape, beaded cuff
AQL	1.5 for holes (1,000 ml water test)
Sterilization Method	Gamma irradiation

PHOTOS OF EVALUATED FABRICS

The following photos show the fabrics of coverall Japan, coverall US, and cleanroom underwear new, washed or washed/sterilized 25 times and 50 times. Photos have been taken through a microscope at two magnifications (x40 and x200). The whole scale is 0.001 m; the distance between two lines is 10 micron.

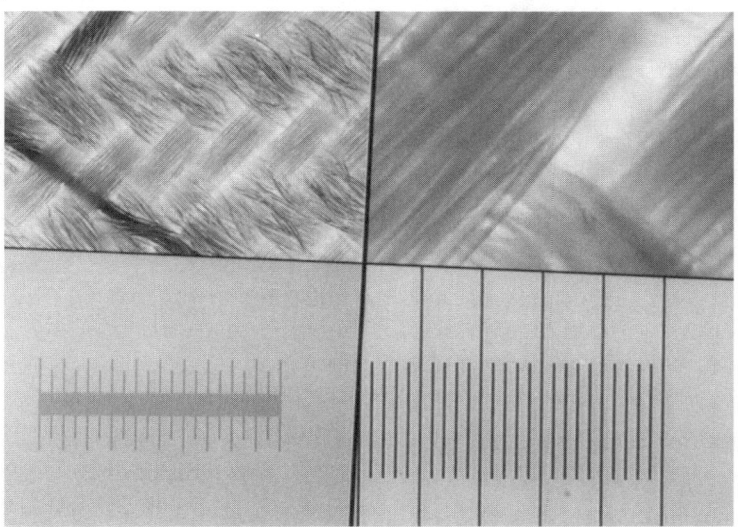

Coverall Japan; new

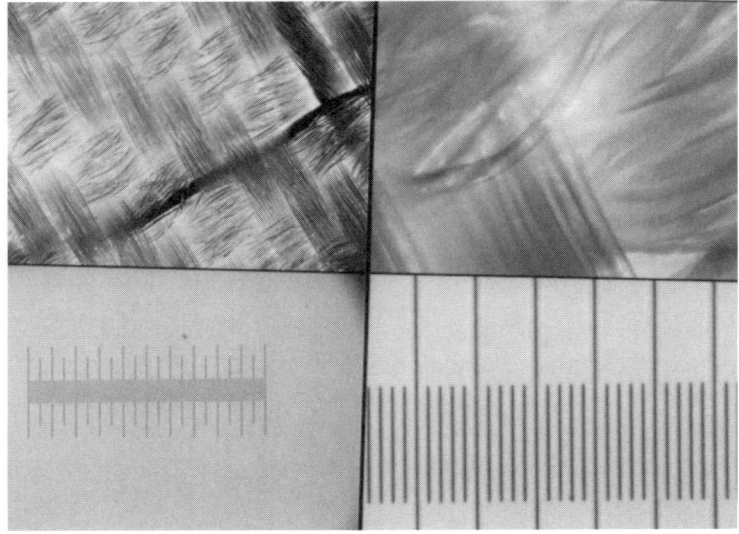

Coverall Japan; washed and sterilized 25 times

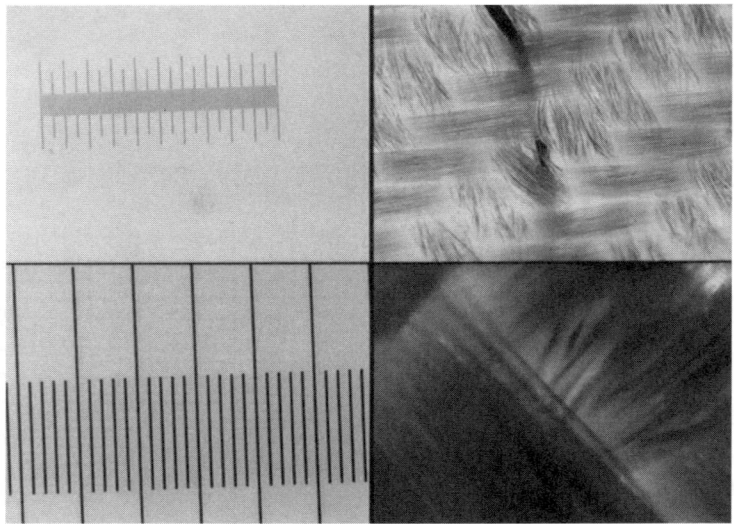

Coverall Japan; washed and sterilized 50 times

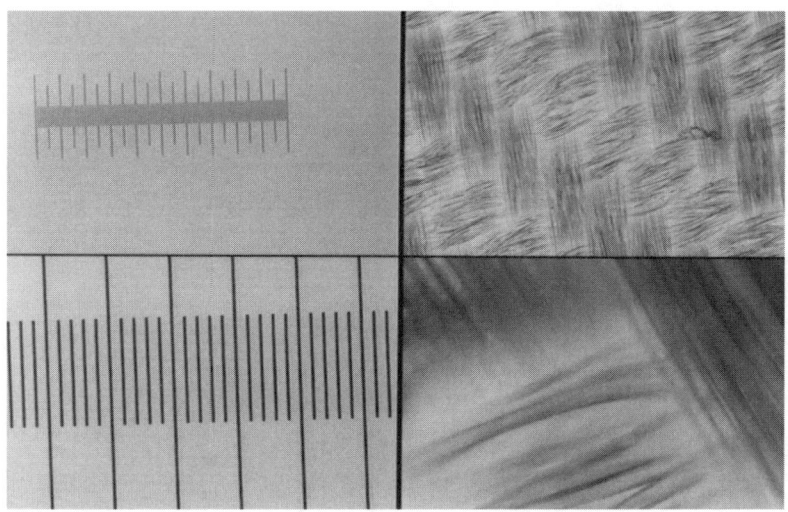

Coverall Japan; washed 50 times

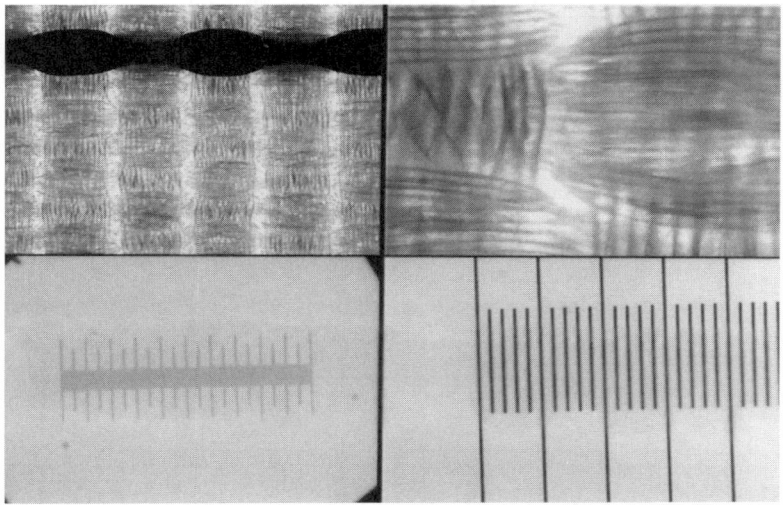

Coverall US; new

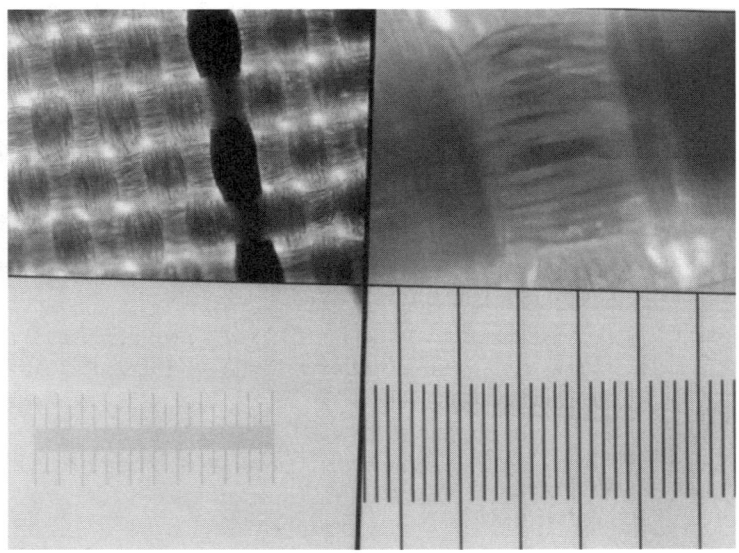

Coverall US; washed and sterilized 25 times

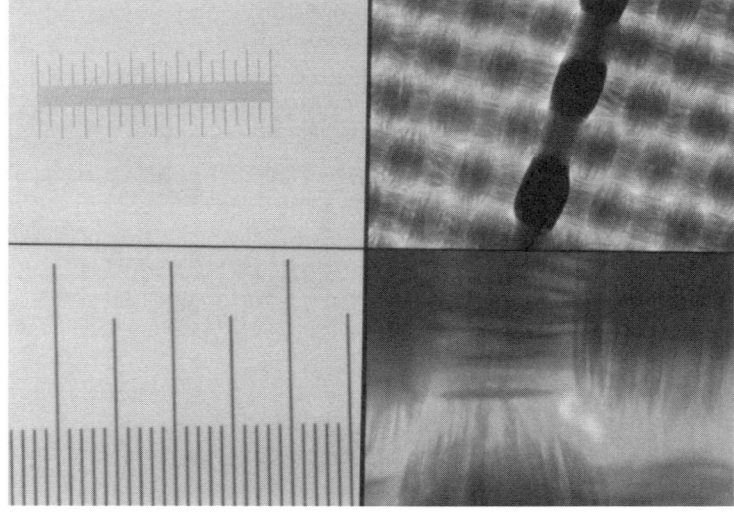

Coverall US; washed and sterilized 50 times

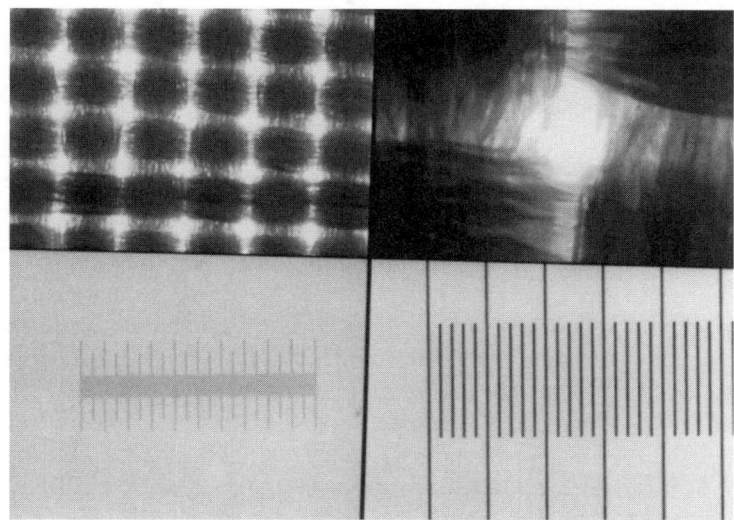

Clean room underwear; new

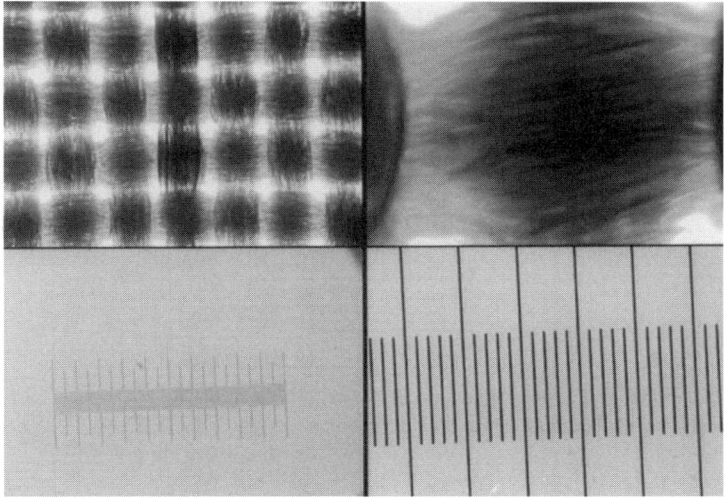

Clean room underwear; washed 25 times

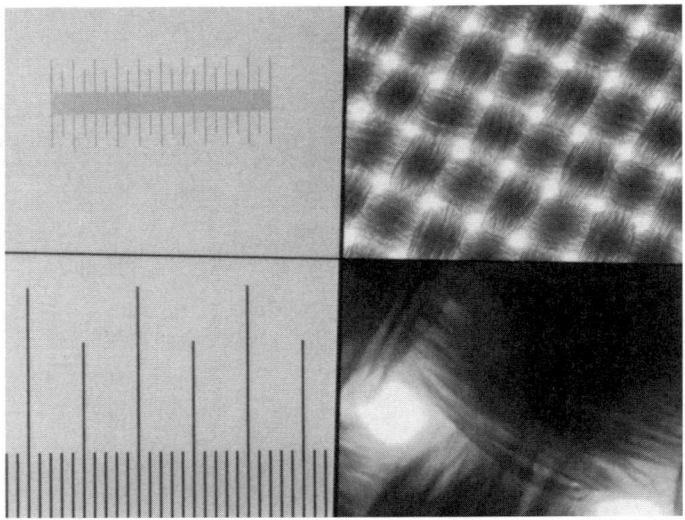

Clean room underwear; washed 50 times

INDEX